WHAT YOUR COLLEAGUES ARE SAYING . . .

"Are you teaching operations, fractions or functions? If so, Harris has some gorgeous ideas for you—showing us the ways they are all 'figure-out-able' with mathematical reasoning."

Jo Boaler
Nomellini-Olivier Professor of Education, Stanford University
Stanford, CA

"*Developing Mathematical Reasoning* is every teacher's guide to breaking away from algorithmic-centered teaching. From the three distortions of mathematics to the hierarchies of mathematical reasoning, Harris helps us understand how math and math teaching have become entangled in a tension between algorithms and reasoning, and then shows us how to untangle this tension through a series of real classroom examples. In so doing, Harris shows us that math is, actually, 'figure-out-able.'"

Peter Liljedahl
Professor of Mathematics Education, Simon Fraser University
Vancouver, Canada

"Harris explores the limitations of an algorithm-centered classroom and emphasizes the need for true mathematical reasoning. By presenting a hierarchy of reasoning domains and advocating for a strategy-centered approach, this book equips educators with vital tools to empower students and deepen their understanding of mathematics."

Graham Fletcher
Math Specialist
Atlanta, GA

"Chock full of real stories about real people engaging with real math, *Developing Mathematical Reasoning* lives up to its title. Harris beautifully empowers educators with practical insights and steps to help students become true mathematical thinkers, not just mimickers—essential for a world that needs confident reasoners."

James Tanton
The Global Math Project
Paradise Valley, AZ

"This is a timely and, ultimately, brave book about mathematics. Harris shines a light on ineffective practices and reminds us that math is so much more than memorized procedures. Her insights may ruffle some feathers about long held beliefs on math instruction. But the invitation to reach more deeply into real mathematics will open many eyes."

John R. Tapper
CEO & Founder, All Learners Network
Burlington, VT

"This book is a must-have! I grew up in the trap of the algorithm. I made it through school with good math grades because I was a good rule follower. It wasn't until I was getting my master's degree that I learned I didn't know mathematics, I was just good at arithmetic."

Christina Tondevold
The Recovering Traditionalist
Orofino, ID

"Harris is not only a dear friend but also an incredible advocate for teaching math in a way that truly empowers students. In this book, she beautifully abstracts the essence of math, guiding teachers on how to help students deeply understand concepts rather than just memorize procedures. Harris has always been a brilliant resource for educators, helping them uncover the 'why' behind the math, and this book is a testament to her passion and expertise."

India White
TEDxSpeaker, Author, National Ed Consultants
Brooksville, FL

Developing Mathematical Reasoning makes the bold assertion that math instruction should teach students to think mathematically. In this day and age when quick answers are coming quicker than ever, Pamela Weber Harris encourages us to slow down. Using concrete examples and vignettes, Harris demonstrates how traditional teaching methods tend to short-change development by pushing procedural thinking. This book teaches how to navigate around those traps and build classrooms rich with reasoning.

David Woodward
Founder and President, Forefront Education
Boulder, CO

Developing

MATHEMATICAL REASONING

Avoiding the Trap of Algorithms

Grades K–12

PAMELA WEBER HARRIS

CAMERON HARRIS, Contributing Writer

CORWIN **Mathematics**

FOR INFORMATION:

Corwin

A SAGE Company

2455 Teller Road

Thousand Oaks, California 91320

(800) 233-9936

www.corwin.com

SAGE Publications Ltd.

1 Oliver's Yard

55 City Road

London EC1Y 1SP

United Kingdom

SAGE Publications India Pvt. Ltd.

Unit No 323-333, Third Floor, F-Block

International Trade Tower Nehru Place

New Delhi 110 019

India

SAGE Publications Asia-Pacific Pte. Ltd.

18 Cross Street #10-10/11/12

China Square Central

Singapore 048423

Vice President and
  Editorial Director: Monica Eckman

Senior Acquisitions Editor,
  STEM: Debbie Hardin

Senior Editorial Assistant: Nyle De Leon

Production Editor: Tori Mirsadjadi

Copy Editor: Talia Greenberg

Typesetter: C&M Digitals (P) Ltd.

Proofreader: Wendy Jo Dymond

Indexer: Integra

Cover Designer: Gail Buschman

Marketing Manager: Margaret O'Connor

Printed in the United States of America

Paperback ISBN 978-1-0719-4826-2

This book is printed on acid-free paper.

25 26 27 28 29 10 9 8 7 6 5 4 3 2 1

Contents

Note From the Publisher: The author has provided video and web content throughout the book that is available to you through QR (quick response) codes. To read a QR code, you must have a smartphone or tablet with a camera. We recommend that you download a QR code reader app that is made specifically for your phone or tablet brand.

Videos may also be accessed at **mathisfigureoutable.com/dmrbook**

Preface: Math Is Math. Or at Least It Should Be.

Have you ever had a moment when your understanding of something shifted dramatically? More specifically, when you didn't realize there was even a different way to think? Have you ever rewatched a film or television show for children as an adult, and realized a line or a character or a story means something completely different to you now?

I had that experience with math.

Math is a tricky thing to talk about.

Partially because math is not supposed to be tricky to talk about.

Math is the universal language. The words we use might be different across times and cultures, but the relationships, the equations—those stay the same. Math is supposed to be the unchanging bedrock of science and technology, the one thing everyone can agree on. In any language, $2 + 2 = 4$ is as uncontroversial a statement as it is possible to make.

The language we use to describe math has changed dramatically over the course of human history. The shift from a system like Roman numerals to our modern base ten place system, the invention of zero, irrational, and imaginary numbers, have all dramatically transformed mathematics. Modern mathematicians are still exploring better ways, looking for the next big jump in the way we describe and do math. And that's all beyond the scope of this conversation.

As uncounted numbers of frustrated parents have exclaimed, "You can't change math! Math is math!"

Unfortunately, that is only correct if what we learned as math *actually was math.*

What if math is actually something different than what many of us thought? What if it has actually been obscured, misrepresented, and hidden behind shortcuts and tricks until it is practically surrounded by half-truths, overgeneralizations, accidental lies, and unconnected trivia, such that our perspective of math can be so damaged we don't understand what math even is? And then we teach the next generation and pass on those misconceptions, and they teach the next, and so on. It's like a bad game of telephone, except the original message was wrong to begin with.

I've found that this miscommunication spiral has resulted broadly in three distinct distortions that affect how many people view mathematics and the teaching of it. The result is that even though there are very different ways people all over the world perceive math, many don't realize there is another way, let alone multiple ways.

Note that these distortions are not binary. Through their life experiences some people are dealing with more distorted math than others, while still others might be dealing with very little distortion at all. However, my experience in 40 years of teaching suggests that the vast majority of people, students or not, are dealing profoundly with one of the following three.

THE THREE DISTORTIONS

THE FIRST DISTORTION: MATH IS NOT FIGURE-OUT-ABLE, IT'S ROTE-MEMORIZABLE

"Are these the problems we did on Tuesday, or the ones from Wednesday?"

"Is this where we cross multiply and divide or find a common denominator? Or cross cancel?"

"I've got this far. What's the next step?"

"6 times 8, like the garden gate, is made of sticks so it's 56. . . . Wait. . . ."

"If I put my fingers up like this, then it's these fingers and those, so 9 × 7 is 6 and 3, 63."

"Does slope go here in the formula? Do I have the right formula?"

"Since math is about memorizing all of these things, I will help students memorize them with mnemonics, stories, and songs."

Under this first distortion, mathematics is an arbitrary set of rules and procedures (such as algorithms). To do well, students must decide what to do, in the right order, and copy the teacher's examples. The why, the background, or the connection to other mathematics doesn't make a difference in getting right answers, so that discussion is irrelevant. "Please just tell me what to do and let me get my homework done." This distortion holds that math has little to nothing to do with your life's experiences.

Some of you read this and nod, "Yes, that's what math is. If my students do this, they will be successful. If they don't have good memories, they won't do well. If they don't do well, they might have math anxiety, they won't pursue STEM fields. That's just how the world works."

This was me.

To be clear, I was *good* at this conception of math. I didn't become a high school math teacher because I struggled with the subject. I *excelled* at thinking about math this way. I *excelled* at teaching it.

Then I found a better way. To my delight (once I got over the existential crisis), I discovered that students struggling with memory can be reached. That students who appear unwilling to exert themselves will self-motivate if the distortions are removed. That students not struggling to mimic can dive deeper and soar higher.

THE SECOND DISTORTION: MATH IS FIGURE-OUT-ABLE FOR ME, BUT NOT FOR EVERYONE

"I mean, you could do all of those steps, but 99 + 47 is just like 100 + 46."

"Why would you go to all that effort? Plainly, $\frac{1}{2}$ of $\frac{3}{5}$ is just $\frac{1.5}{5}$, so $\frac{3}{10}$."

"Why do I have to show those steps? It's just obvious."

"I'm not sure why no one else is just figuring these out. Maybe they need the steps."

"I guess I have the math gene. I don't know how to help people without the math gene think like I do."

"I do things in my head, but I know I'm supposed to teach the rules and steps. That's what you do."

Under the second distortion, math is figure-out-able, but for some reason not for everyone. Math-ing means to use what they know to reason about new things. Someone who is under the spell of the second distortion believes that "I can add to my repertoire and keep building because it all makes sense. I do not have to wait until someone shows me a rule. But for some reason, other people can't or won't."

Many with this distortion were taught by well-meaning people with the first distortion. And they thought that teaching was ridiculous, nonsensically inefficient, or at best not needed. They watched teachers conduct sing-alongs and teach rhymes to memorize algorithms, wondering all the time what in the world any of this had to do with math.

This was my eldest son's experience, who upon being taught a subtraction algorithm in first grade thought it was unnecessarily complicated and invented his own. This is using $7 \times 7 = 49$ to reason that $8 \times 7 = 49 + 7 = 56$. This is reasoning about the equation of a new function using the equation of another function without starting from scratch with a formula.

To fit the definition of this distortion, this reasoning ability is gained *in spite* of how someone was taught, not *because* of it.

The problems this distortion cause become most obvious when it is time for this person to turn around and *teach* math. They learned real math, but didn't realize it was in spite of how they were taught, not because of it. They teach their students the same way they were taught. Some students "get it" the way they did, but most don't. Many assume at this point the difference must be that of innate ability, that the students who get it have "the math gene," and those without, don't.

Sometimes when I discuss these three distortions, people say, "I want to have the second distortion." But remember, this is a distortion because people under this distortion do not realize they can purposefully teach what they actually did naturally while their teacher was drilling step-by-step procedures. Of course, we all wish we had the natural talent to recognize mathematical patterns without being intentionally taught them, but we can't invent natural talent. What we can do is teach the real math-ing to students instead of the fake math of memorizing and mimicking.

THE THIRD DISTORTION: MATH IS FIGURE-OUT-ABLE, BUT NOT FOR ME

"I still don't understand. I think I should be able to understand, but this seems really random."

"Yes, I could do what you're telling me. But no, I don't want to just do the steps because it doesn't make sense."

"I mean, I could just try to memorize and do what you're saying, but I know I'll mix it up because I don't get it."

"Math is hard to understand. I'll do my best to clearly explain the parts I get and be patient to explain as many times as needed."

These people think math should make sense. They should be able to figure out what to do because memorizing what is arbitrary doesn't work for them. They have a sense that *math* is not arbitrary but, for whatever reason, don't make the connections the same way people under the second distortion do. Some students under this third distortion believe their teachers are deliberately holding back, deliberately not explaining the math.

Because math is so often presented as something to rote-memorize, these students are left to figure it out on their own. These students often abandon their reasoning when it doesn't match what the teacher is doing, assuming that reasoning must be wrong. The algorithms work against their intuition, invalidating their thinking.

Rote-memorize: to commit a fact to memory independent of the surrounding context that explains why that fact is true. For example, memorizing the names of capital cities with flash cards or memorizing multiplication tables without building the accompanying Multiplicative Reasoning that explains and justifies why the multiplication table is the way it is.

Math starts looking like a bad magic trick. Any piece of math—subtraction, the Pythagorean Theorem, pi, the slope-intercept equation—all might as well be ink blots to memorize, because none of them make any sense.

The rest of the book will illustrate how algorithms work against intuition.

These students are stymied when $7 \times 8 = 56$ makes as much internal sense as pineapple times automobile equals tiger. They try to memorize the songs and pictures and sayings, but they know they're not relevant. Knowing $7 \times 7 = 49$ is of no help learning $7 \times 8 = 56$, because there is no connection between pineapple times automobile equals tiger and pineapple times airplane equals lion.

Many, many people I've talked to about "memorizing their multiplication tables" saw them exactly this ridiculously. Disconnected and meaningless.

Given that reality, many under this distortion disengage from the learning as a defense mechanism. They invest less emotionally, to soften the repeated shaming that not understanding brings. They succumb to "Just tell me how to do it; I'll fail at that, and we can move on."

The truth is that often their teachers are operating under one of these three distortions. Individuals working under the first distortion don't know math is figure-out-able, and so understandably won't teach it that way. Those under the second distortion know math is figure-out-able, but don't know how to teach it that way. Finally, those dealing with the third distortion never felt like they succeeded learning math themselves. They teach the way they were taught, hoping their students will figure it out where they didn't. Usually while stressing about it. A lot.

TRY IT

Consider which of these distortions resonates with you or with your experience as a student. Consider how the lens you've had may have colored your view of mathematics or mathematics teaching.

THE REALITY: MATH IS FIGURE-OUT-ABLE FOR EVERYONE

Real math, math-ing, is not trivial. It is not obvious. It is not simple. But it can be taught.

In my 30 years in math education, the one truth that has been reinforced over and over again is that everyone can do more real math than fake math. Everyone can do more math when that math is built on what they already know rather than shoehorned on the backs of contextless rote-memorization. In other words, everyone can *math*. Everyone can have their horizons open up and have more choices.

What does it mean to "math"? Real math, doing real mathematics, begs a verb like math-ing. Deborah Crayton has coined the term math-er. "Readers read. Writers write. Mathers math" (Crayton, 2026). Cathy Fosnot uses mathematizing (Fosnot & Dolk, 2001, p. 4). Math-ing or mathematizing as a verb describes the mental actions that mathematicians do. See Chapter 1 for more about what this means.

The first distortion is inherently limiting. The mountain of facts and steps to memorize become too much. Learners can't keep it all straight or use any of it to reason about new things. For many, this happens as they move into long division, fractions, or algebra.

People under the second distortion usually make it the farthest. But how much more could they have learned faster if their growth were assisted by their teacher, instead of having to figure it out on their own? How many more people could join them in these STEM fields if they were actually taught real math-ing?

When helped to *math* in a real way, people who were under the third distortion gain the confidence to invest emotionally again because their effort is rewarded. Frustration and anxiety vanish into comprehension and proficiency. They know they can understand, and indeed they do.

In university classes and in-service workshops I lead, when I get people math-ing, many for the first time, I get these reactions:

- First distortion: Whoa, I did that. That was *my* thinking. Wait, we can teach math this way?

- Second distortion: Yes, that's what I've been doing in my head, but now I'm seeing that I can teach kids to do what I've been doing. Cool.

- Third distortion: Hallelujah—I knew I could understand! Now I can help my kids *math* with understanding too.

TRY IT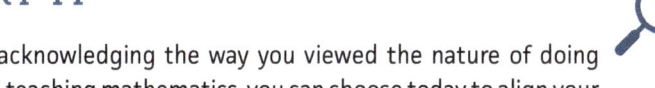

By acknowledging the way you viewed the nature of doing and teaching mathematics, you can choose today to align your teaching with what you actually believe.

Take this quiz online:

https://qrs.ly/x1g41d1

To read a QR code, you must have a smartphone or tablet with a camera. We recommend that you download a QR code reader app that is made specifically for your phone or tablet brand.

It is reproduced on the next page.

The Perspective Quiz

What did you think it means to do and teach mathematics?

When you think about learning mathematics as a child, do you . . .

A. Break out in a sweat, get nervous, and wonder if you'll remember how to do the problems?

B. Smile, remembering some fun problems you worked on and patterns you found that helped you make sense of problems?

C. Feel cheated, like you know you could have learned more and done better if your teachers would have explained more, or better?

D. Remember knowing that if you just practiced a lot, you could remember what to do when?

If your childhood friend had asked you a mathematics question, you would have . . .

A. Clearly told them the rule and the steps to do the problems or looked it up in a textbook or online help.

B. Told them they will have to ask someone else because you never did understand how to do those problems.

C. Looked at the problems to see what relationships you could use to solve them, but then showed your friend the steps you learned in school.

If you had missed a day of school, you would have . . .

A. Looked at the missed assignment, confident that you could probably figure out the answers to the problems by thinking logically about them.

B. Waited until the teacher showed you the rule and the steps to solve the problems.

C. Tried to figure out how to solve the problems, but if you couldn't readily, asked the teacher to explain what was happening and why, so that you could understand how to solve the problems.

If you didn't understand a teacher's explanation, you thought that . . .

A. It might be your fault, but maybe you just were not a math person.

B. You could sit with the problems, think about them, and figure out a way to make sense of them.

C. You needed to see the steps again and practice some more.

When a teacher began to explain the lesson for the day, you hoped . . .

A. The teacher would help you understand what it all meant and why because you knew then you had a chance of getting it right. If you didn't understand, you might get correct answers today, but you wouldn't be able to hang on to them without understanding.

B. The teacher would just tell you what to do and how. "Give me the steps and let me practice them. Please don't tell me why or give me more than one way."

C. You'd have a chance to play with the concept, the numbers. You wanted to try your hand at solving the problems on your own. If the teacher made you mimic their steps but you didn't need to, that was frustrating.

A teacher said, "Show me your work." What you heard was . . .

A. Use what you know, how you understand what's going on, to make sense of the problem. Then write something on paper, probably what the teacher had shown, because you may not know how to write down what was happening in your head.

B. Copy exactly the steps that you were shown in the right order. Practicing the correct steps in the correct order is the work— that's what it means to do mathematics.

C. Show me that you understand what the teacher was asking. If you did not understand, you didn't want to just do what the teacher had shown because then you knew you wouldn't be able to do it again. It didn't make enough sense for you to own it.

FREQUENTLY ASKED QUESTIONS

Q: What if none of these distortions feels like they apply to me?

A: First, remember that these descriptions are generalizations and not meant to describe any one person exactly. Second, many people feel like they've changed at some point. They feel like they started out believing that math made sense and that they were capable of reasoning through it and then later found themselves sty-mied. For example, *This year's math isn't making sense like last year's*

(Continued)

(Continued)

did, or *I thought I was a math person, but I guess not.* Many students begin their schooling with a clearer idea of what math is than they have by the end of their first multiplication/fraction/long division unit. They then end up under the first and second distortions. You could use this discussion to understand why this global conversation about mathematics education is so complicated, tricky, and subtle, because many are coming from these different distortions. The most important part of this discussion is to point us all to real mathing. The three distortions are frameworks that suggest why people might disagree about how mathematics should be best taught.

Q: What if my teachers taught me conceptually, and now I teach conceptually too? I don't seem to fit your three scenarios.

A: Give the rest of this book a read. If you get to the end, and sure enough, your teachers and you approach teaching mathematics as developing a hierarchy of reasonings that don't rely on mimicking any algorithms *and* that every single one of your students can learn math this way—then, fantastic. You are one of the lucky few. Otherwise, you might be dealing with the second distortion.

ABOUT THIS BOOK

Chapter 1 lays out how the implementation of algorithm-centered math education today, its methodology and goals, are often at cross purposes with the true nature of mathematics and doing mathematics. It then goes deeper into how the proliferation of algorithm-centered teaching is largely responsible for these issues, and that understanding those issues presents the best opportunity for improving math education. When we understand the nature of mathematics, we can mentor students to *math* like mathematicians.

Chapter 2 introduces the hierarchy of mathematical reasonings essential to the learning and progress of all mathematics students. It goes through the major domains of reasoning: Counting, Additive, Multiplicative, Proportional, and Functional, laying out how they build off each other and represent tiers of increasingly sophisticated thinking. It introduces *sophistication* as a descriptor of the thought processes used when solving math problems. This term is needed because although it includes ideas of speed and efficiency, it also includes the magnitude and complexity of mathematical relationships in use, which neither speed nor efficiency denote. The chapter will go into this topic in far more depth, but I want to note here that the term *sophistication* as it is used in this book is never a value judgment of a person or their thoughts. It is only used as a relative measuring stick to place where a given method of solving a problem falls on the growth continuum. A student currently developing in the Counting Strategies domain of reasoning is not better or more valuable than a student developing Functional Reasoning. They are simply at different stages of development.

Chapters 3–6 each define and illustrate one reasoning domain and how rote-memorizing and mimicking algorithms can trap students into using less sophisticated reasoning than the problems call for, therefore limiting students' reasoning growth. Each includes a detailed, step-by-step walkthrough of at least one commonly used algorithm and an explanation of how at each step it can undermine students' opportunity to grow their mathematical reasoning ability. These chapters illustrate the major mathematical strategies to develop in place of those algorithms and discuss the advantages the strategy-centered approach brings. These advantages include an often faster, almost always longer lasting, and more complete understanding of content.

As used in this book, the term *mathematical reasoning* does not mean just a general ability to think. This is not a fuzzy, "think better" approach that doesn't include doing the math and getting results. Mathematical reasoning is about building stronger brains and expects more, not less, from students, giving them the tools to actually be successful at math-ing. It demands increasing sophistication of strategy. This means meeting students where they are, and then helping them develop from there. For example, students will not only know their multiplication facts, they will actually own them and be able to use the relationships in problems. It includes content-specific milestones such as understanding of integer addition and subtraction, multiplication of fractions, and so forth.

The final chapter, Chapter 7, answers the question, If mathematics teaching is not all about repeating the steps of algorithms, then what is it? The chapter outlines steps teachers can take to improve their own and their students' mathematical reasoning ability regardless of their current reasoning level or what content they need to teach.

For the content you teach, you can work to solve problems using what you know and learn the major models and strategies for that content. You can work to elicit and represent student thinking, making thinking visible, point-at-able, and discussable. Lastly, you can work on high-leverage teacher moves and sequencing tasks, with an eye toward moving the math forward and meeting all students' needs.

Each chapter includes tips and FAQs throughout, as well as actions the reader can take—either personal exercises or things to try in class.

Corwin and I will be publishing four additional grade-specific companion books (K–2, 3–5, 6–8, and 9–12) on a six-month cadence once this book is released, which will offer more ideas, more practice, and more practical advice, concentrated specifically on each grade band. These books will be complementary to this anchor volume, which we believe is necessary to set the foundation of the discussion on developing mathematical reasoning.

FOUNDATIONS

All of the ideas, concepts, methods, and proposals for how to teach more students more math contained in this book have their foundation in 30 years of study and classroom-based research. *Development of Mathematical Reasoning* is the result of synthesizing research and personal experimentation with teachers and students in real classrooms to find what works and what doesn't—what cultivates real understanding versus what gets quick answers at the cost of long-term development.

My work is influenced by that of Fosnot and Dolk (2001) in their *Young Mathematicians at Work* series and Fosnot's *Contexts for Learning*, which showed me children reasoning about content and how to get them to do it; Jean Piaget (1896–1980), the founder of cognitive development; Hans Freudenthal and the Freudenthal Institute in the Netherlands and their Realistic Mathematics Education philosophy; Constance Kamii and Ann Dominick, who published "The Harmful Effects of Algorithms" in 1998; and Liping Ma, who coined the PUFM "profound understanding of fundamental mathematics" in her *Knowing and Teaching Elementary Mathematics* (2010). Other noteworthy influences are the work of Marilyn Burns, *Math Recovery*; Les Steffe, Anderson Norton, Amy Hackenberg, and Susan Lamon's *Teaching Fractions and Ratios for Understanding*; NCTM's *The Teaching and Learning of Algorithms in School Mathematics* (1998); *Developing Mathematical Reasoning in Grades K–12* (1999); Kazemi and Hintz, in their Intentional Talk (2014); Smith and Stein (2011) in their *Five Practices for Orchestrating Productive Mathematics Discussions* the textbook *Functions Modeling Change* by Connally et al. (2000); and recently *Building Thinking Classrooms* by Liljedahl (2021) and *Rethinking Disability and Mathematics* by Lambert (2024).

We already have many of the foundational elements of teaching math better. The last four major standard shifts in the United States—"Professional Standards for Teaching Mathematics" (NCTM, 1991), "Principles and Standards for

School Mathematics" (NCTM, 2000), the "Curriculum Focal Points" (NCTM, 2006), and the Common Core State Standards (2010) are important moments in recent history that each tried to delineate what should be taught when in school mathematics. The move toward developmental progressions based on research was necessary and helpful.

Simultaneously, the National Science Foundation funded several universities to create textbooks that were more aligned with that current thinking. Textbook series like *Discovering Mathematics; Investigations in Data, Number, & Space; Math in Context;* CMP (the Connected Math Project), *Everyday Mathematics, Math in Context, CORE Plus,* and COMAP had many schools and teachers trying to get students to investigate and discover math that was more in context, using manipulatives, models, and technology. Many teachers, who were like the earlier me, tried these innovative approaches but didn't understand why it was necessary or what the goal even was. Most importantly, the new standards and the textbooks based on them do not account for the heavy distortions that most teachers and students operate under about the nature of math—for example, the myth that one must have the math gene to excel at math or that math must be memorized and mimicked.

This book brings together the outstanding research that exists and the understanding of the way it has been misunderstood to help leaders and teachers navigate where to go now.

A word about research.

It is frankly *easy* to take two groups of students for a few weeks, drill one group and not the other, and then show that the drilled group "knows" more. Research like this rarely gives any indication of how long students will retain what was drilled, or if the isolated drilled "knowledge" is weaving well (or at all) into the interconnected web of mathematical knowledge needed to further support learning. I find this research unhelpful.

The research that I find much more useful consists of those studies where researchers create tasks, facilitate them with students, learn more about how students learn and the interconnectedness of the mathematics, tweak based on the results, share what they have learned, and then rinse and repeat. These more useful studies show what students know long after the initial teaching and how it connects to and supports future learning. This research helps me as a mathematician, a mathematics teacher educator, and as a teacher myself to better understand *mathematics for teaching* (Ball, Thames, & Phelps, 2008), and how to help teachers and students develop as genuine doers of mathematics.

HOW TO USE THIS BOOK

This book is meant to be read from beginning to end, at least on the first read-through. You may be tempted to skip to what looks like your grade level, but *an essential part of avoiding the traps of algorithms is understanding the prior context of what comes before your grade level.* Once you've gotten to your grade level, don't stop there: Understanding how your content affects reasoning in the latter grades is also essential to developing mathematical reasoning.

There are frequent problem-solving examples throughout the book. When you reach one, pause! Think. Solve. Think about your thinking. Examining your own thought processes and pondering how your students would react to these problem-solving opportunities is a crucial part of making the most of this book. Fundamental to the Math Is Figure-Out-Able philosophy is that we have to *math* to learn how to teach math better. You must establish the relationship in your own head so you have something to hang other people's thinking on.

After a first reading, teachers seeking to hone their skills and understanding about their specific subject areas would do well to study the chapters covering those topics. Coaches or leaders may want to read the whole book multiple times—first from a learner's, then a teacher's, then a leader's perspective.

In this book, I use teacher's and student's names where I have permission and pseudonyms where I do not. I am so grateful for the expert teachers who allowed me to work with them and their students.

Discussion Questions

1. When you think of math as a verb, *math-ing*, do you think more of reasoning, creating arguments, justifying, critiquing, or do you think more of rote-memorizing and mimicking?

2. Do you recall being under any of the three distortions? What sparks for you when you read these descriptions? Is anything missing? How would you tweak them?

3. Might your colleagues be under any of these distortions? How do you know?

4. What are you wondering about as you finish reading this preface?

Acknowledgments

To my son Cameron, a mighty thanks for being the kid who blew my mind over and over again, giving me the gift of real math-ing. And how fun is it that you helped me write this book?! Without you, I don't think I would have gotten this done. Definitely not with as much humor.

Thanks to my other children, Matthew, Craig, and Abigail, who were the best guinea pigs, and my husband, Daniel, for everything, including helping me make the sticky bits less sticky. And Abby—great job on the graphics!

Thank you to Kim, Sue, and the rest of the gang at Math Is Figure-Out-Able. And thanks to Ann Latham for her meticulous work with permissions.

Thank you to Garland Linkenhoger, the Charles A Dana Center at the University of Texas, T3-Teachers Teaching with Technology, Paul Kennedy, Cathy Fosnot, Scott Hendrickson, Kathy Hale, and all my colleagues who have shaped my teaching career.

Thank you to Debbie Hardin, Senior Acquisitions Editor at Corwin, who has walked me through the whole book-writing process with a skilled hand. And thank you to everyone at Corwin for giving me the opportunity to get this message to the world.

Here's a respectful and grateful thank you to the thousands of teachers who have been with me on my journey, taken our workshops—online and in person—and allowed me into your classrooms. You work tirelessly to improve the lives of your students. It shows. I hope this book does justice to your journey.

And to the Author of life, thank you God for giving me a message worth sharing.

PUBLISHER'S ACKNOWLEDGMENTS

Corwin gratefully acknowledges the contributions of the following reviewers:

Brandon Pelter
Math Teacher, Bridgeport Public Schools
Norwalk, CT

Kimberly Rimbey
Chief Learning Officer and CEO, KP Mathematics &
Kim Rimbey Consulting
Author, Corwin
Phoenix, AZ

Adina Rochkind
5–8 Math Curriculum Coordinator
Baltimore, MD

Brendan Scribner
Math Consultant, Scrib Mather, LLC
White River Junction, VT

Laura Vizdos Tomas
Math Coach, School District of Palm Beach County
Cofounder, LearningThroughMath.com
West Palm Beach, FL

About the Author

Pamela Weber Harris is changing the way we view and teach mathematics. Pam is the author of several books, including the *Numeracy Problems Strings K–5* series, *Building Powerful Numeracy,* and the *Foundations for Strategies* series. As a mom, a former high school math teacher, university lecturer, and an author, she believes everyone can do more math when it is based in reasoning rather than rote-memorizing or mimicking. Pam has created online *Building Powerful Mathematics* workshops and presents frequently at national and international conferences. Her particular interests include teaching real math, building powerful numeracy, sequencing Rich Tasks to construct mathematics, using technology appropriately, and facilitating smart assessment and vertical connectivity in curricula in schools PK–12. Pam helps leaders and teachers make the shift that supports students to learn real math because math is figure-out-able!

CHAPTER 1

Math Is Figure-Out-Able

A nyone like gum?" I ask a class of 28 high school seniors. We are about to video classroom interactions to put in my *Developing Mathematical Reasoning* online workshop. This is the warm-up the day before, to let me get to know the students, practice pronouncing names correctly, and give students a chance to know what to expect when the film crew arrives the next day.

I start with a *Problem String*, an instructional routine designed to build a specific mathematical model, strategy, or concept.

These students are taking a course called Advanced Quantitative Reasoning, an alternative to calculus. This likely means these students had been fairly successful in their previous courses, algebra 2 and precalculus, but did not choose to take calculus for their last year of high school math.

"If 1 pack of gum has 27 sticks, how many sticks are in 10 packs?" I ask.

I don't wait very long for this answer. Maddie replies, "270 sticks."

"How about 9 packs? How many sticks in 9 packs?" I ask.

After some think time, Kayla replies, "It's 243 sticks. Just take away 27 sticks from 270."

"How did you do that subtraction?" I ask.

"Take away 30 and give back 3," she says. I represent that strategy on the board, and we briefly discuss it and two other ways students were reasoning about 270 – 27.

I follow with, "How many sticks in 100 packs?" and Victor answers quickly with 2700 sticks.

When I ask, "What about 99 packs?" students smile. Not a common experience for many teachers when asking 99 times anything.

Mia answers, "Just take away 1 pack. So, 2700 – 27. That's 2673."

After a couple of students share how they reasoned through the subtraction 2700 – 27, Cameron, sitting right in the middle of the class, looking very thoughtful, raises his hand.

"It's almost"—Cameron pauses—"it's almost like you want us to use *what we know* to solve these problems."

"Yes," I smile. "Yes, I do."

Every other student in the room nods thoughtfully, like this is new and noteworthy. A realization that is both wonderful and tragic. Wonderful, because these students have just experienced what it means to really do math. Tragic, because experiences like this should have defined their entire math education, not just be featured as a last-minute footnote.

"Keep using what you know," I continue. "What if we had 2646 sticks of gum—how many packs?"

Students look at what is being represented on the board and use that to reason that 2646 is 54 sticks, or 2 packs, less than 100 packs. These students have confidently reasoned that 2646 sticks are in 98 packs, or $2646 \div 27 = 98$, without steps, without mimicking.

What students are doing is not a bunch of random mental math tricks. These students are developing one of the major strategies that spans the operations and makes use of the distributive property—the Over strategy. In this case they are multiplying by a bit too much and adjusting back.

It's focused, purposeful instruction in real math-ing.

Is this what math class looked like and felt like when you were a student?

I am not advocating that we stop teaching grade-level content, making seniors back up to multiplication in isolation. This particular Problem String is so good because we can get students reasoning about multiplication, work on Additive Reasoning with subtraction, and simultaneously build Proportional Reasoning because we're using a ratio table as a model. We can then extend that String to graphing those ordered pairs, writing the function to match the data, $y = 27x$, because the number of sticks equals 27 sticks per pack times the number of packs. We can also discuss the graphs of related transformed functions, like the line containing (1, 28) and (2, 55) or the line containing (1, 26) and (2, 53). There is a lot of meat here, real content with many access points to allow students to enter the problems and also to challenge all students. You can find more examples of Problem Strings that build numeracy into middle and high school content in my Building Powerful Numeracy for Middle and High School Students *(Harris, 2011) and* Lessons & Activities for Building Powerful Numeracy *(Harris, 2014). Also watch for grade-specific companion volumes I am publishing with Corwin (K–2, 3–5, 6–8, and 9–12), which I am publishing in order every six months, starting with the release of this anchor volume.*

MATH IS ACTUALLY FIGURE-OUT-ABLE

Worthwhile mathematics teaching is about helping students to use what they know to reason through problems, strengthening their minds as they grapple with and make sense of increasingly complex mathematics.

The purpose of mathematics instruction is to build students who *math* (Crayton, 2026), not students who are mimicking or randomly guessing. Teaching mathematics, mentoring students to *mathematize* (Fosnot & Dolk, 2001a), requires focused direction and pedagogical skill that capitalize on students' existing mathematical knowledge and intuition and guide them to develop ever more sophisticated powers of reasoning.

Math is not rote-memorizable; math is not random-guessable. Math is figure-out-able.

SWIMMING WITHOUT WATER

Many current math classes operate like learning to swim without water.

To learn to swim, a person gets in the water. They begin by doggy paddling when they can just barely reach the bottom. As they learn strokes, become more confident, and build stamina, they swim farther and in deeper water. In many math classes, it's as if students are watching from outside the pool, observing people swim and mimicking what they can see above the water, but they actually have no idea what's going on under the water. Students are not privy to the thinking going on in a mathematician's mind, the choices made along the way, the start and stop, try, fail, and correct that happens when doing something new. Students are told they're swimming as they sit on the edge with feet dangling in the water, but they never actually learn to swim.

When students in math classes rote-memorize facts in isolation and mimic procedures, they're being told they've learned more and more math, but in reality their brains never get any stronger mathematically. Their mathematical reasoning ability isn't growing, only their catalog of memorized facts. These facts are essential, but they are not sufficient.

Mathematical reasoning in this book is not some generalized problem-solving schema, some fuzzy thinking better. Mathematical reasoning includes content—this means that a student reasoning mathematically is using mathematical relationships, properties, and models to argue, operate, and solve problems and in the process learn more mathematics content. You will learn more about this in Chapter 2.

Students are told they are doing math as they mimic procedures, carry the 1, cross out the 0, keep change, flip. But they aren't encouraged to see the beauty of a well-formed argument, a clever strategy, or a model that illuminates and helps them own interconnected relationships. When math students don't know what it means to critique reasoning, logically prove generalizations, and use what they know using mathematical relationships, they are not experiencing real math-ing.

Many of us were trained to teach mathematics as rote-memorizing steps and mimicking actions without ever engaging in the mental actions that mathematicians use. This is the way most of us learned, ourselves. But as Maya Angelou said, "When we know better, we do better." In the sticks of gum Problem String, those students were using what they knew and reasoning with mathematical relationships, not parroting procedures. Procedural mimicry is not doing real math—it's swimming without water. Many of us have been trapped over the years into thinking that we are doing the work of mathematics, when in reality we aren't math-ing at all. *Mimicking squelches opportunities to develop more sophisticated thinking, the doing of real math.*

> ## TRY IT
>
> Consider ways that you have inadvertently done the heavy lifting on your students' behalf, or allowed them to push down the metaphorical piano keys but kept them from doing the real work of math.

To be clear, I am not blaming individual teachers or schools. I am *certainly* not blaming you. For heaven's sake, I was the teacher who was trying to get students to swim when I didn't even know what swimming *was*. I used rhymes and sound effects (literally) when my students could have been producing actual music. This book is my attempt to help you learn what I did, to have the epiphanies I did, without the years of research and experimentation it took me.

This isn't about blame or shame. It's about helping you do and teach more real math.

When I say math like a mathematician, I don't mean the math-ing that mathematicians do as full-fledged adult mathematicians; I mean the way mathematicians mathematized in first grade, fifth grade, eighth grade. Over lunch in the dining hall in the Queen's College at

OUR BIGGEST OPPORTUNITY FOR IMPROVEMENT

Many math education articles, books, and talks begin with a montage of depressing statistics about math achievement—more specifically, the lack thereof, the dwindling pool of students with the math qualifications to follow STEM careers, and so on. Most conclude with some flavor of insistence that this woeful situation is because our math students' speed is not speedy enough, and their accuracy is not accurate enough, then propose ways to fix those problems. Many of the proposed solutions describe where teaching should fall on the continuum between extreme versions of direct teaching and inquiry. One extreme proposes demonstrating to students exactly how to solve all the types of problems they will need to solve in a particular class. The other argues that we should give kids prompts and manipulatives and let them explore and discover how to solve problems on their own.

Some blame direct teaching for poor student performance: "Stop doing that," they say. "Do discovery, inquiry learning instead. It's more engaging, fun, and real world when students do problem-based learning. Get kids some conceptual understanding so they'll be able to do the algorithms better." These teachers might hope that the right combination of exploration and engagement will help students understand and learn the algorithm better.

Others claim the opposite, that not enough direct instruction is happening in the classroom: "Stop trying to get them to discover everything—that is fuzzy math, where teachers are making students guess at and reinvent math. Why make kids struggle? Just tell them how to do it clearly. Have kids memorize the basic steps so they can build the more complex math on that foundation." Teachers on the far end of this spectrum might teach math the same way they would teach capital cities in a geography class—with flash cards, rhymes, and mnemonic tricks, as if there was no underlying logic behind the answers. They I-do-we-do-you-do their way through a chapter in a textbook, demonstrating algorithms and watching to make sure students can do all the steps correctly.

These two approaches feel like opposites, but they share the same misguided goal: that students should get answers easily and quickly and that the best, most efficient way to do that is by repeating algorithms. Procedural fluency, understood as easily performing the steps, is the hallmark. Either way, repeating the steps of an often inefficient, nearly always opaque, mind-numbing algorithm is put on a pedestal as the supreme example of what mathematics competency looks like. Where they differ is how to teach those algorithms.

I am using terms to describe some muddy philosophies: direct, explicit teaching and discovery, inquiry teaching. To make my arguments clear, I am caricaturing both, describing the extreme ends of the spectrum. The reality is that most often the argument between the two is an unhelpful false dichotomy, particularly when algorithms are the goal. I argue we need to change the goal. Then the whole conversation shifts.

Let's follow this to its logical conclusion. If the goal of math class is to produce students who produce correct answers, mathematics education could be reduced to a few weeks covering how to use generative AI.

Math class in the 21st century cannot be about answer-getting (Daro, 2014).

Once we recognize that producing answer-getters cannot be the point of math class, teaching algorithms loses much of its justification. If we change the goal of math class, algorithms as teaching tools are no longer the focus.

Our greatest opportunity for improvement lies in removing algorithms from the goal of math class. Believing that doing math means memorizing and mimicking algorithms, even with understanding, is the same as believing that writing the steps of a workout would ever be sufficient for getting in shape.

That watching someone riding a bike will provide the experiential balance and coordination necessary to actually ride the bike yourself.

Our greatest opportunity for improvement lies in removing algorithms from the goal of math class.

Removing algorithms as teaching tools in math education will not automatically fix all of mathematics education, any more than removing a splinter will heal a wounded foot. But just as removing the splinter is necessary for the healing to begin, removing algorithm-repetition from the goal of mathematics class will shift the conversation about teaching in a way that will help us refocus on what matters most: helping students learn to think and reason mathematically

FREQUENTLY ASKED QUESTIONS

Q: But, Pam, I learned all of those math-y things *because* I learned the algorithms. When I use an algorithm, I am not mimicking, I'm thinking mathematically. We need to teach those algorithms because that is how I learned the math I know and learned to think the way I do.

A: This is a very tricky conversation. There are subtle things at play here. Would you consider . . . Is it possible that you had natural inclinations to pick up on patterns and relationships, so that when your teacher showed you an algorithm, you created many mental connections with things in and around that algorithm? And since you did, you might now associate all of those self-made connections with learning that algorithm. The mathematical connections like place-value, magnitude, rounding, and friendly numbers might be inextricably linked (in your experience) to the steps of an addition algorithm and the experience of learning it because you were thinking through the steps and making sense of it using your natural talent.

Might you be willing to consider that all of those extra mathematical connections had less to do with being shown steps to repeat and more to do with your natural proclivity to pick up on and use patterns? And if that's true, might it be possible that you could have learned far more, faster, if someone had been actively, purposefully helping you develop those connections? If you did it all on your own, imagine what you could have done with purposeful, expert guidance.

(Continued)

(Continued)

Mathematicians of old and the random nonmathematician ran into the same patterns in life but the mathematician noticed, used, built on them. The random nonmathematician did not. You were able to do what you did, recognize patterns and make connections when presented with an opaque algorithm, without that expert guidance. You know that most students can't do what you did with the same (lack of) support. Many of us could not—we bought into what we were told: memorize and mimic. And even if we tried to make sense of it (pick me), we didn't have the natural talent to do it without those patterns being made explicit.

The question is not whether or not algorithms work as teaching tools for most of the population. They just don't. The question is whether or not we can teach most of the population *and* still give advantaged students like yourself what they got from learning through algorithms.

The answer is that we can, and we can give students like yourself so much more. And the best way to teach it is to high-dose everyone with those patterns.

Q: What if algorithms are required in our standards?

A: Great question! I won't pretend this is trivial, but there are solutions. You can meet your standards and teach real math-ing. This will become clearer throughout the book. Keep reading, and we'll wrap it up in Chapter 7.

BEING TRAPPED BY ALGORITHMS

I want to tell you about a few people I know who were trapped by a math education that did not focus on empowering students to use what they know to figure out what they don't know yet. This misdirection in math education almost always stems from an addiction to algorithms as teaching tools.

At one point in graduate school, I was a teaching assistant for a large college algebra class. The 300+ class met three times a week in a large hall, and students could visit me during my office hours for help with homework and studying for tests. Most of the students who came to me were elementary education majors. They would lament, "I hate this class! Please help me get through it so I'll never have to take math again." They had been trapped by an exclusively speed-and-accuracy-focused math education.

Years later, elementary and middle school education majors in my math methods courses would complete a math autobiography as part of their first assignment. Over 90% of my students would introduce themselves as wanting to teach kindergarten, first grade, or maybe second grade because they did not believe they could teach fractions. They had been trapped into believing they could only teach young grades because of their anxiety around mathematics.

One of my goals every semester is to help my students leave my class ready and excited to teach the grade level they actually want to teach, not just the one they previously thought was the highest math they could teach. Each semester it's one of the best things that happens when students happily, confidently report seeking and obtaining those positions.

Kyle Pearce, cohost of the *Making Math Moments That Matter* podcast and cofounder of the Make Math Moments company, has told the story before that he was a university student when he realized that he didn't really understand mathematics. His mathematics education had trapped him into thinking he was good at math and he was severely disappointed to find that his preparation was so insufficient. His powers of memorization were good, but he hadn't developed the complexity of mathematical thought his professor was looking for. Kyle is now a top-notch coach, math teacher educator, and task designer (Pearce & Orr, 2020).

I myself hit a wall hard in advanced university math courses. I had more than a decade of being rewarded for superb speed and accuracy, only to have my Abstract Algebra professor tell me, "Oh, we don't do that here," when I asked him to give me the proofs ahead of time: "If you'll just tell me what proofs will be on the quizzes and tests, I will memorize them and spit them back out on the assessments perfectly." He just shook his head. I had been trapped by excellent grades for mimicking algorithms into believing I was doing real math when in reality I had been succeeding at rote-memorizing and mimicking, not math-ing at all. I had not developed the sophistication of thinking I needed and wanted.

These are just a few examples of people who have been trapped by algorithm-focused math instruction. We'd memorized piles of algorithms, but instead of climbing higher with each one, we were drowning in more and more disconnected facts and procedures. We'd learned, as the Dutch mathematician Hans Freudenthal said, that school mathematics is like the "fossilized remains" of real mathematics (Freudenthal, 1973). In being taught this way, we'd gained, if anything, the perverse ability to get answers without any of the foundation required to make use of them.

WHAT IS AN ALGORITHM?

So let's make sure we're on the same page with vocabulary. An *algorithm* is defined as a series of steps to solve any problem of a particular kind. It is the same method for every problem, regardless of the numbers or structure. All the steps, every time (Carpenter et al., 1998; Kamii & Dominick, 1998; Plunkett, 1979; wolframalpha.com, 2024).

This definition of an algorithm is almost universal in science, statistics, computer science. Over the past few years in mathematics education, some people have started to use the word algorithm *when* strategy *would be more precise. This muddies the water. See page 23 for more on the distinction between algorithms and strategies.*

Algorithms are a general solution, which means an algorithm can solve even the gnarliest of problems. Algorithms are often opaque, where the place-value, magnitudes, and meaning are hidden behind the scenes, cleverly embedded so the user does not have to deal with the complexities that are occurring. This is the beauty, cleverness, and remarkableness of the algorithms. It is also what makes them terrible teaching tools.

The algorithms I refer to as the traditional algorithms and use as examples in Chapters 3–6 all share the commonality that they can be counted on to reduce any problem of a type to single-digit arithmetic. For example, the traditional North American multiplication algorithm will turn any multiplication problem, no matter how many digits are involved, into a series of single-digit multiplication and single-digit addition problems (see Chapter 4). This is very powerful, but also inherently limiting.

Memorizing one algorithm is usually no help with memorizing the next one. Knowing the steps of a traditional multiplication algorithm will not help with memorizing the steps of the traditional long division algorithm.

This means that for many students math class becomes increasingly frustrating and difficult to manage. As they progress through the grades, what they learned last year does not help them understand what they are learning this year.

Because students do not actually have to understand what is going on to perform the steps, students can use less complex reasoning than those problems could help develop. "Fantastic!" critics cry. "Students will be able to do so much math without really having to deal with the complexities (i.e., learn anything).

This is a desired outcome! More students doing more math. Who wouldn't be in favor of that?"

But students using algorithms are only *mimicking* more math. As Liljedahl (2021) observed in his book *Building Thinking Classrooms*, "Everywhere I went I saw the same thing—students not thinking and teachers planning their teaching on the assumption that students either couldn't or wouldn't think" (p. 12).

Our goal in teaching mathematics cannot be to make things *easy*, particularly if it means sacrificing long-term growth for short-term answers on homework and test scores. If that was acceptable, we could hand out the answer key and call it a day. We as teachers know this instinctively. Some amount of struggle is necessary for learning. But that does not mean all struggle is useful or created equal. Struggling to memorize and mimic is effort spent *now* for more confusion *later*. Grappling with and making sense of real math pays huge future dividends. Real math is knowledge that builds on itself.

TRY IT

Think about the algorithms that you teach your students. Having that list in the forefront of your mind will be useful as you continue to read.

Our world needs thinkers and reasoners, so our world needs a math class that trains thinkers and reasoners. Liljedahl (2021), who encourages us to *Build Thinking Classrooms*, said, "My goal from the outset was to get students to think. . . . Thinking is a necessary precursor to learning, and if students are not thinking they are not learning" (p. 296).

That is why I object to using algorithms as teaching tools. As Hurst and Huntley (2018) found, "most students in [their] sample are 'prisoners to procedures and processes' irrespective of whether or not they understand the mathematics behind the algorithms."

Algorithms are amazingly powerful mathematical inventions that are essential to much of modern technology. The issue is that the skill of developing and finding uses for algorithms is vastly different from the skill required to follow the steps of an individual algorithm. The former can program a computer to land a rocket on the moon; the latter can be replaced by that 50-year-old computer landing that rocket on the moon.

The ability of a human mind to create an algorithm is amazing; the ability of a computer to execute an algorithm is revolutionary; and the damage done by using algorithms to teach mathematics is incalculable.

FREQUENTLY ASKED QUESTIONS

Q: But mathematicians use algorithms, right? So if we want our students doing math like mathematicians, our students should be using algorithms, right?

A: In a study where mathematicians were given problems to solve, they used an algorithm only 4% of the time. That means that 96% of the time mathematicians reasoned through the problems, using relationships they know. Mathematicians create algorithms, study algorithms, compare algorithms. They don't use them to compute (Dowker, 1992).

Q: What if my standards require algorithms?

A: The short answer is that in most circumstances you can meet your standards and avoid the trap of algorithms. The longer answer is in Chapter 7 and will make more sense after reading Chapters 2–6.

Q: Why do you refer to the *standard* algorithms as *traditional* algorithms?

A: There is nothing standard about the algorithms that have become traditional. The word *standard* denotes "one and only" and gives too much weight and credibility. People are often shocked to find there are several different algorithms commonly in use around the world. My mother grew up in Switzerland and does division completely differently than what I learned. The subtraction algorithm my eldest son independently created in second grade is the same as the method taught in many Latin and South American countries.

CONFUSING LOGICAL-MATHEMATICAL KNOWLEDGE FOR SOCIAL KNOWLEDGE

By now you are probably wondering why teaching algorithms evolved as *the* way to solve problems in math class. To answer this question, let's parse out the difference between social/conventional knowledge and logical-mathematical knowledge.

Child development psychologist Jean Piaget suggested there are three types of knowledge (Piaget, 1974):

1. **Physical knowledge:** The understanding of the physical world.
2. **Logical-mathematical knowledge:** The understanding of being able to solve problems and perform analytical reasoning.
3. **Social knowledge:** The understanding of societal norms and conventions.

The last two types are important when we discuss learning math: logical-mathematical knowledge and social knowledge.

The trouble in math education is that we have a history of treating mathematics as all social knowledge. In reality, most of mathematics is logical knowledge. Next, let's make sense of the difference and how they both are best taught.

Historically, mathematics was only for wealthy individuals who could afford expensive educations. Throughout history, math-y people had developed mental relationships that culminated in really cool general algorithms. As education democratized, schools tried to bring math to everyone. Schools did so with textbooks that lifted those algorithms from that body of work and handed them to teachers and students as if the algorithms are all the math there is to learn and not one small application of it. The resulting misalignment of scope resulted in the limiting idea that answers are the best evidence of learning math. That has been passed down, and over time it became the picture of what math is. Even today, as people write scope and sequences, lessons, and textbooks, they tend to cut up the interconnected web of relationships into tiny memorizable pieces, making mathematical ideas into falsely linear sequences, giving all things equal weight, and making it all about answer-getting. That put us very far away from what math-ing actually is.

Before, we just didn't know how to pull back the curtains and help students develop mathematical reasoning. Now, we do. To learn how, read on!

SOCIAL KNOWLEDGE

Social knowledge is that which we deem to be so.

Another way to say this is that someone suggested social knowledge as knowledge, and over time *by convention we all adhere to that idea.* In math, there is a small set of things that must be told—they cannot be figured out. This set consists mostly of vocabulary and notation that history has tapped as *the* way to say it or write it. Given that these conventions are not figure-out-able, students would only be able to guess, and more often than not they would guess incorrectly.

For example, we cannot ask students to reason to find the name of a four-sided polygon (many angled figure). If they use patterns, 10-sided is decagon, six-sided is hexagon, five-sided is pentagon, so surely a four-sided polygon is called a quadagon? A fouragon? Squareagon? Oh, actually we have the tradition of calling them quadrilaterals (four-sided figures). We can't reason our way through tradition. Good grief! What is a three-sided polygon called? A tri-ilateral? A thrice-agon? (See Figure 1.1.)

FIGURE 1.1 ● Social Knowledge Versus Possible Figure-Out-Able Names for Common Shapes

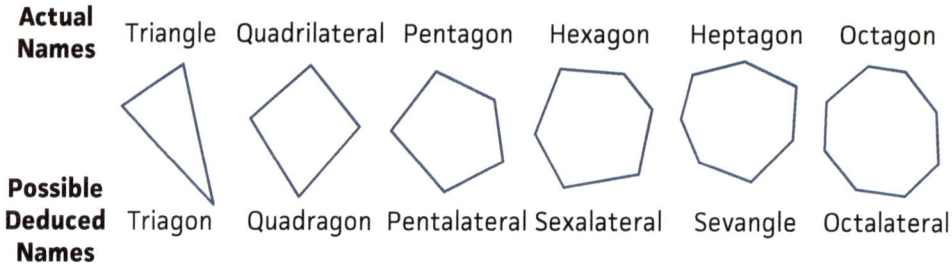

We can't figure out what the notation is supposed to mean without some social help. Students could guess, but they'd likely guess incorrectly because there is no way to logically reason or use what they know.

Notation is another example of social knowledge. For example, parentheses mean multiplication, 3(4) = 12, except when they don't:

- $f(x)$,
- $f^{-1}(x)$
- the point (2, 3)
- interval notation (−∞, −2)

We can't figure out what the notation is supposed to mean without some social help. Students could guess, but they'd likely guess incorrectly because there is no way to logically reason or use what they know.

Let's not make students guess about the parts of mathematics that are social knowledge. This set of knowledge is the things we should purposefully tell students with intentionality and straightforward clarity.

TIP

Teach vocabulary just in time, not just in case. When you teach vocabulary just in case, students learn lists of terms and definitions beforehand, in case they will need them. Rather, teach vocabulary just in time. Give students experiences where they are begging for a way to describe what they are thinking about and dealing with. Teaching just in time means hanging terms on already constructed logical-mathematical knowledge.

LOGICAL-MATHEMATICAL KNOWLEDGE

Most of mathematics, however, is *logical-mathematical knowledge*—that which must be experienced, connected, reasoned, *figured out* in order to actually learn, own, and use meaningfully. To reason mathematically, these things cannot simply be told, memorized, or mimicked. The best way to teach students mathematics is to teach them to mathematize, to do the kinds of thinking involved in math-ing.

This logical-mathematical knowledge is the power to see 99×675 and, rather than funneling into an algorithm of steps to parrot, instead use Multiplicative Reasoning to recognize that 99 groups is one less than 100 groups, and therefore the answer is one less 675 than 67500.

Such connections are everywhere in mathematics, even in areas you might think right now can't possibly be figured out. We know *this bit of math* because someone figured it out in the first place, then created the algorithm we might think is the only way to solve it. We know this because such connections define what mathematics *is*.

Are there parts of 99×675 that are social knowledge? Yes, the look of the numerals and the multiplication symbol are social knowledge, and that is a fruitful teacher discussion—the parts of math that must be told and the parts that must be experienced and worked through with logic. I invite you to consider that the set of things that are social knowledge in mathematics is *far smaller* than we have been led to believe.

This conversation is in part difficult because so many of us were taught math as *all social knowledge*: Wait to be told what to do, then rote-memorize and mimic *everything*.

If we are teaching a part of mathematics with a mnemonic, rhyme, or story, we are treating it as social knowledge— that which must be told and memorized.

UNPACKING THE CONFUSION

Learning the names of rivers that traverse a country or all 50 capital cities in the United States is social knowledge. These things must be rote-memorized.

What about multiplication facts? Is something like 7×4 social knowledge? Pause here. Take inventory of what you think. Are multiplication facts something to be clearly told and then rote-memorized (social)? Or are they to be connected and reasoned about (logical-mathematical)?

Multiplication facts are not like the random names of polygons, rivers, or capital cities. They are internally consistent and logically built from each other. Eight groups of eight, 8×8, is 64, and one more group of eight, 9×8, is 72, and $64 + 8 = 72$. But a student might never realize those connections if they are instructed to rote-memorize each fact. Imagine what a student thinks about the facts if they "learn" the facts as is suggested in a "learning" video program I encountered on YouTube. I've summarized a part of it for you as follows:

> Learn 7×4 in under 10 minutes! All you have to do is memorize this story. Mrs. Week (7) sits on a chair (4). She goes fishing and catches 2 boots and 8 fish. You got it: $7 \times 4 = 28$.

The video even instructs students to not remember the story wrong—she didn't catch 8 fish and 2 boots. Suggesting that 2 boots and 8 fish is not the same as 8 fish and 2 boots sends the message that this is all arbitrary and that addition is not commutative. It also sends the message that 28 is made up of 2 and 8, not 20 and 8. This is a prime example of sacrificing your future multiplicative reasoning self for a current third-grade story repeater. Sure, some kids will rote-memorize facts more easily with a story. But then that's all we get—students repeating nonsensical stories pretending they are doing math. Consider the chance students have to *understand* fractions if they believe multiplication facts are arbitrary, disconnected, random *vocabulary*. They are trapped.

To help explain the prevalent obsession with rote-memorizing multiplication facts, consider this. When many of us were learners in the midst of performing the multiplication algorithm over and over again, it became painfully clear that crunching through each of those problems was easier if you had the single-digit facts at your fingertips. This striking and strong memory leads us to think, "To help students, let's make sure that they rote-memorize all of those facts." In our embodied memory of our student experience, we didn't need to understand more, so an alternative was not even on our radar.

Of course, we want students to know their multiplication facts! But the mathematical reality is that students need to *more* than know them. Students need to travel the mental path of figuring the facts often so that those paths become well traveled. Once we can get students reasoning using the connections between multiplication facts, they learn those facts *and* the ones those facts connect to *and* they develop Multiplicative Reasoning at the same time. "Providing students with opportunities to think about things differently, find similarities and differences, and evaluate which approach is best, all enhance the brain's construction of new learning" (Jensen & McConchie, 2020, p. 175). Multiplicative Reasoning is the goal. As we develop Multiplicative Reasoning, owning the facts becomes a natural by-product.

Multiplication facts are one of the first opportunities to help students develop Multiplicative Reasoning. Multiplicative Reasoning looks like finding 9×8 by thinking about $10 \times 8 = 80$, but since that's too much, removing one 8, for 72. This helps with thinking about things like 49×6 by thinking about $50 \times 6 = 300$, but since that's too much, removing one 6, for 294. "This kind of deep dive into the learning ensures the brain understands the wider or deeper context and not just isolated or rote facts" (Jensen & McConchie, 2020, p. 175). This reasoning Jensen and McConchie describe is logical-mathematical and essential for everything that comes afterward, like fractions, proportions, functions, etc. Chapter 2 will continue to illuminate how to develop reasoning.

The issue with memorizing multiplication facts the same way you would memorize capital cities in a geography class is that *mathematics class is not a geography class*. Rivers, mountain ranges, capital cities' names are social knowledge. To know these is to memorize. The multiplicative connections and relationships between facts are logical-mathematical knowledge. To know these is to *math*.

HOW TO TEACH

If everything in mathematics were social, then we would clearly need to tell students all of it.

This just isn't the case.

Most mathematics is not social knowledge. And because most mathematics is not social knowledge, it is unhelpful to give procedures and hope students guess at the reasoning underlying them when it is the reasoning that is important. Students will get answers, but they probably won't build reasoning.

We need to approach the teaching of mathematics with the recognition that most of the material is logical, adheres to patterns, and can be deduced from other knowledge that is already understood. This does not mean that students should do all of the connecting on their own—I am certainly not an advocate for unfocused, anything goes, fumbling, fuzzy math teaching.

While these essential, extremely powerful connections are everywhere in mathematics, that does not mean they are so obvious that students should be left to their own devices to discover them. I'm not advocating that we turn classrooms into directionless vacuums, where students are left to their own devices to forage for math.

Students benefit from a "more knowledgeable other" (Vygotsky, 1978) in a classroom where teachers are the guides who craft experiences, give feedback, mentor, and support every student mathematician. Students need teachers who assess what fledgling knowledge students already have and help students build on that prior understanding. Not as a minor item on a checklist before a lesson but as the foundation on which the entire classroom experience is built. They need teachers who know the mathematical relationships and how to help students develop these relationships and connections for themselves.

Kim Montague, my pithy cohost on the Math Is Figure-Out-Able *podcast, gives us the challenge to "Know your content, know your kids." (Montague, 2021) This is not just a side-note, catchy phrase. Knowing your content and knowing your students are the bedrocks on which we build the foundation of our math classes. Most of this book is to help you build your content. Equally important are knowing your kids, assessing what they know, how they learn, what motivates them, their culture, and their interests. Giving your students open enough tasks so they all have access, and they are all challenged in appropriate ways, is the exciting and important work of building on the content with which they come to you.*

Efforts at discovery-based teaching frequently end up with kids doing the best they can alone trying to guess what's in the teacher's head. That can feel callous, and it is not helpful. Often that triggers a reaction of frustration from parents and colleagues: *Why hold back on students? Just tell them the things. Don't make them reinvent math—just tell them what to do in the clearest possible way. Give them plenty of practice to make that mimicking stick.* As teachers we might be lulled into a false sense of security, believing that if students are getting answers, and parents are happy, everything is fine. We might also believe that all math is social knowledge. Therefore: tell, memorize, mimic, practice mimicking.

Except, as we've just discussed, everyone is not fine when students are rote-memorizing and mimicking, fake math-ing. Even the "successful" students quickly find that the "math" they've learned is nearly useless outside a fake math classroom. Students' mimicking muscles are getting a workout, but students' mathematical reasoning abilities are staying stagnant or atrophying. Many of us—teachers and parents—are unknowingly perpetuating a fake math myth.

If we clearly understand the parts of mathematics that are social versus those that are logical-mathematical, deciding how to teach each part is simple: Construct most of mathematics by mathematizing with students, giving them experiences so their brains can make the connections. Tell only the small set that is social knowledge (see Figure 1.2).

FIGURE 1.2 ● A Decision Tree to Inform Teaching Practices

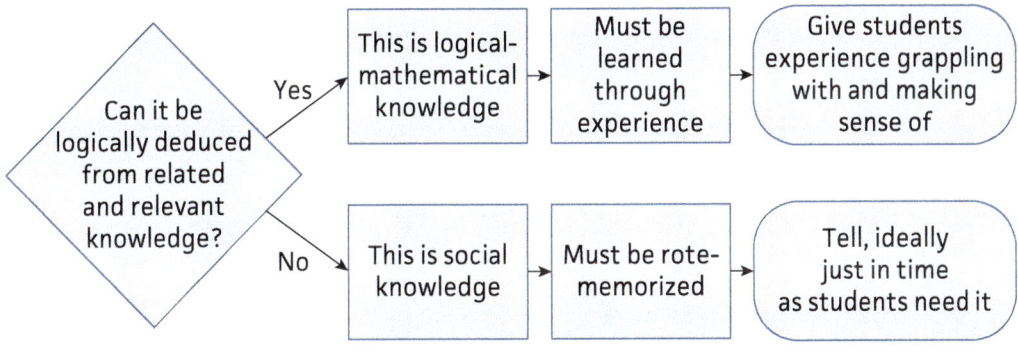

It is real work to learn to differentiate between what is social and what is logical, between the arbitrary words and notation we've chosen and the real depth and expanse of mathematics. Between what is window-dressing we only need for communicating with each other and what is figure-out-able.

But that's why you are reading this book, right?

WHAT IS MATHEMATIZING, MATH-ING?

What does it look like and feel like to mathematize, to math the way a mathematician *maths*?

Fosnot and Dolk (2001a, 2001b, 2002, 2010) in their Young Mathematicians at Work *series have a chapter titled "Mathematics or Mathematizing?" in which they discuss the purpose of mathematics education. And Crayton (2026) is one of those who has made math into a verb: math-ing. Freudenthal suggested, "What humans have to learn is not mathematics as a closed system, but rather as an activity, the process of mathematizing reality and if possible even that of mathematizing mathematics" (Freudenthal, 1968, p. 7).*

Here's a glimpse into mathematizing: What is 99 plus anything? What is 99 times anything?

Think about the problem 99 + 47. How do you reason about that sum?

- Could you think about 100 + 47? Add a bit too much, so adjust one back?

- Could you think about *one more* plus *one less*, 100 + 46?

We can use both of those strategies to reason about 99 plus *anything*.

Think about the problem 99 × 47. How do you reason about that product?

- Could you think about 100 × 47? That's too much, so adjust one group back?

We can use this *over* strategy to reason about 99 times *anything*.

These are examples of using what you know, adding 100 or multiplying by 100, to reason through a problem. This is the work of mathematizing, of math-ing. This kind of work strengthens your brain and builds your capacity to deal with more complex ideas and relationships (see Figure 1.3).

TIP

When you see problems like these in the book, solve them before you read on. Getting a sense of the relationships involved will help you make sense of the commentary that follows.

FIGURE 1.3 ● The Work of Math-ing

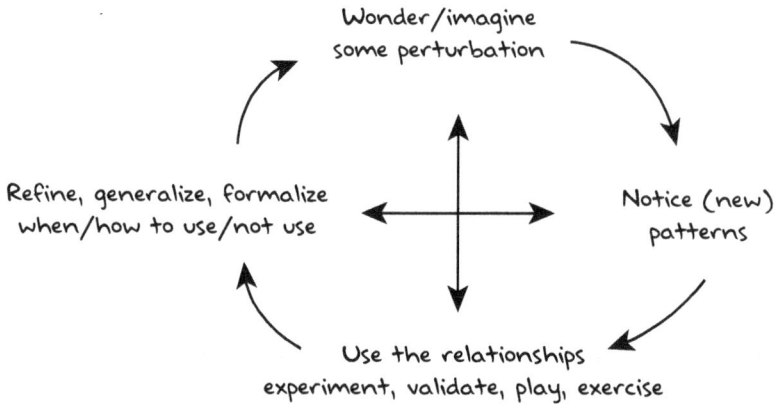

Mathematics teaching at its best means to give students something mathematically important to think about, where they can wonder and imagine. This leads to helping students notice patterns that students can use to experiment, validate, play, and exercise their fledgling ideas. With expert teaching help, they can then refine, generalize, and formalize their thinking about when and how to use or not use the relationships, which leads to the opportunity to wonder about new things on their horizon.

We can mentor students to mathematize by helping them develop such mathematical strategies.

FREQUENTLY ASKED QUESTIONS

Q: You just used the easy number 99. Of course, you can reason with such easy numbers. What about all the other numbers? Don't we need algorithms for those numbers?

A: No. For any number or problem that is so complicated that you might need to reach for an algorithm, you could instead reach for a calculator. The major mathematical strategies that I outline in this book give students access to *enough* problems. That's why these strategies were chosen: to build mathematical reasoning and to give students power over *enough* problems. For more discussion on problems that are worth solving without a calculator, see Chapter 7.

WHAT IS A STRATEGY?

We just discussed a strategy of adding or multiplying by 99. But what is a *strategy*? Using a strategy means letting the numbers or structure in a problem influence how you solve it. Your strategy is how you use the relationships you already know to solve a problem. It means you don't use the same method for every problem because the numbers are not the same, and that you use only what you need to solve the problem, not someone's prescribed steps (Fosnot & Dolk, 2001a; Fosnot & Dolk, 2001b; Wright et al., 2006). Where algorithms are often opaque (difficult to see what's happening behind the scenes), strategies are transparent because you are using relationships *you know* to reason through the problem.

Mathematical strategies are approaches to problem solving that are distinguished from each other by the underlying understanding of mathematics required to use them. These strategies are the ways that naturally math-y people use the patterns they intuitively notice. Just because the rest of us did not pick up and use the patterns on our own, does not mean that we cannot. We can all benefit from deliberate, higher doses of the patterns.

Unlike algorithms, learning strategies are synergistic. Where each new algorithm to memorize is another series of steps to potentially misremember and confuse with each other—was it keep, change, flip, or keep, change, change?—strategies are mutually reinforcing. The more relationships you own, the more strategies you learn, which in turn builds more relationships. This positive feedback cycle is what learning real math-ing is all about.

For example, there are four major subtraction strategies that represent the mathematical relationships a student needs to own. To be clear, I'm not advocating learning four different algorithms instead of one. Decades of classroom experience have shown that building strategies, by leveraging and developing student intuition instead of destroying it, takes far less effort individually and pays far more dividends than even one algorithm required to solve the same class of problems.

Unlike traditional algorithms, major strategies do not reduce problems to single-digit arithmetic. While this sounds like a weakness, it is actually a massive learning advantage. It enables strategies to take advantage of what students already know beyond the most basic of single-digit operations. Instead of reducing 120 × 9 to a long series of single-digit problems, we can leverage what we've learned in math class since third grade to instead solve 120 × 10 = 1200, and 1200 − 120 = 1080. The former stops building brains the moment single-digit operations are conquered, while the latter lays the foundations for proficiency of fractions and Proportional Reasoning.

FREQUENTLY ASKED QUESTIONS

Q: But traditional algorithms are so efficient and save time, right?

A: If the argument is that algorithms save time because a student uses one way to solve a problem every time, then we're not considering the time spent teaching and practicing these algorithms. We're also not considering the time lost in a year remediating students who are unable to mimic these procedures, the years lost in a student's mathematical journey because they quit math, misunderstanding what mathematics really is. By contrast, strategies are natural outcomes of relationships students *need to own*. They are extensions of the mathematical properties that are at the heart of mathematics. Strategies offer students choice so they are able to consider which relationships they want to work with rather than waste time attempting to remember an algorithm they never understood. Given any problem that's reasonable to solve without a calculator, we can be as efficient as a traditional algorithm or, most of the time, *more* efficient. Many examples of this efficiency of strategies follow in the rest of the book.

Q: It sounds like you're not advocating for direct instruction or inquiry. What are you advocating for?

A: I'm advocating for a shift in goals—from mimicking algorithms to developing mathematical reasoning (which includes content). With that new goal in mind, teaching then becomes: good guided inquiry for everything that is logical and clearly telling for the bit that is social. Teachers have clear goals: help students grapple *long enough*; guide students to important generalizations through purposefully crafted discussions; anchor learning; and keep building on that learning to move the mathematics forward using open access enough tasks that all students continue to have access and continue to be challenged. By doing this, students are not just solving problems correctly and efficiently, but also more sophisticatedly. This allows students to be more successful longer. More on this in Chapter 2.

For any problem that's reasonable to solve without a calculator we can be as efficient as a traditional algorithm or, most of the time, more efficient.

Conclusion

The purpose of math class is to develop mathematical reasoning, not mathematical answer-getting. What we need are not mere calculators but thinkers, do-ers of mathematics to solve problems we have yet to encounter. Our role as teachers is to guide and support students as they develop their mathematical reasoning, not rotely mimic algorithms that only provide answers to existing problems. Knowing the traps of algorithms empowers us to make other choices.

Discussion Questions

1. What is the difference between an algorithm and a strategy? The book will further differentiate these, but what are your current thoughts?

2. What is the difference between logical-mathematical knowledge and social knowledge?

3. What is an example of a bit of mathematics that is social knowledge and therefore told to students? Do you and your colleagues agree on this?

4. How were you taught the multiplication facts, as logical-mathematical or social knowledge? How did that affect your perception of the nature of mathematics?

5. What do you think of *math* as a verb? What does it mean to you *to math*?

TRY IT IN YOUR CLASSROOM

99 Plus Anything

Purpose

The purpose of this short interview is to become more aware of the strategies many people use intuitively. It also gives you the opportunity to practice your questioning and listening skills, and your parsing of people's mathematical thinking. Seek to pull out people's reasoning, teasing out what they mean. Ask, don't tell.

This can be quite challenging if:

- you're used to listening solely for correct answers or correctly mimicking steps

- you're like I was, with the algorithm as the sole method you use to solve problems

- you have yet to try to figure out other people's alternative strategies

Use these interviews to open your horizons in a low-stakes environment. Just have fun!

Routine

- Interview several people (your family, neighbors, community members, students, colleagues, anyone willing).

- Ask, "What is 99 plus anything?"

- If the person is confused, clarify, "What is 99 plus any number?"

- If more clarification is needed, add, "Pick an ugly number." (Smile when they choose a number that ends in 7) and ask, "What's 99 plus your [37]?"

- Listen, watch, ask to hear what's happening in their head.

- Try to repeat back to them what they did, putting your own words to their strategy.

- Try their strategy with a different number and ask them if you understood their thinking.

Important to Consider

Make this a casual conversation, not an interrogation. You don't want students, friends, and family members to feel like they're on the spot, especially if they are in front of their peers. Reassure them by suggesting that you're practicing learning how people think about math-y things when they're not necessarily trying to please a math teacher, the way they would actually reason. Ask clarifying questions until you understand what they are thinking. If they start to tell you about an algorithm, don't make them explain those steps. Instead, gently probe for what they might do without those steps.

Some people may add the tens, add the ones, then add those sums. Others may use an over strategy, finding $37 + 100$ and adjusting back 1. Some people may give and take, taking 1 from the 37 to give to the 99, making an equivalent problem, $36 + 100$.

Extension

Depending on the age or experience of your interviewee, you could ask any of the following:

- 9 plus anything

- Anything minus 9

- Anything minus 99

- 9 times anything

- 99 times anything

- Adding $9\frac{9}{10}$ to anything

CHAPTER 2

Developing Mathematical Reasoning

T he goal of mathematics education is to develop mathematical reasoning.

I'm lying on a table in a doctor's office, living in that wonderful liminal, in-between space between agreeing to an injection and waiting for the needle.

Trying to distract me, the doctor asks, "What do you do?"

"I'm revolutionizing the way we teach math," I answer, watching her put gloves on. "Math is figure-out-able, and teaching it that way is the best way to help all students."

"That's quite the statement," she says with an eyebrow raised. "What do you mean?"

At this point, the nurse is arranging the mobile live X-ray above my shoulder, but pauses and looks up.

I answer, "Instead of having students rote-memorize disconnected facts and mimic procedures, we can actually build on what they know and help them think about math, solve problems in math, the way mathematicians do."

"Huh," the doctor says. "That sounds maybe like how my son is being taught. And I like it. I hated memorization when I was in school. My teachers just told me to memorize the facts, but my son actually understands that 7 times 8 is seven 8s." She pauses, fiddling with something I couldn't see that was probably about to hurt. "But there's a problem."

"Oh?" I ask, gritting my teeth as the slow injection starts.

"Yeah," the doctor continues. "I love that my son seems to actually understand what's going on, but he's so slow at solving problems. Figuring out 3 × 5 by adding 3 groups of 5 seems fine, but he also does all that adding for facts like 8 × 7. He gets really bogged down."

I want to nod, but decide not to risk moving my shoulder. "Makes sense. Adding in groups like that is using Additive Reasoning to solve a multiplication problem. A good place to start, but ideally your son would grow to reason multiplicatively. He would group the groups in cool ways *and* be able to solve multiplication problems more efficiently than he would with an algorithm."

She sounds intrigued. "That's a thing?"

I smile. "Very much so."

Now, setting aside that in that same doctor's appointment they found a bone spur in my shoulder, I would have very much liked to conclude that conversation by pointing that doctor to a lovely book that would explain the domains of mathematical reasoning and how they provide a framework for teaching real math, but at the time no such book existed.

This is that book.

TRY IT

Ask at least 10 teens or adults, "What is 7 × 8?" to determine if they think that it's rote-memorizable or figure-out-able, and how. In the conversation, ask how they know/found it. Listen purposefully to determine if they had rote-memorized it, skip-counted, or used a multiplicative relationship. Did they have a song, rhyme, mnemonic? Did they add seven 8s? Did they use a related multiplication fact to find it? If they answer, "I just know it," wonder aloud how 7 × 8 is related to eight 8s or seven 7s.

SIMULTANEITY: AN INDICATOR OF SOPHISTICATION AND COMPLEXITY

The ability to grapple simultaneously with an increasing quantity of numbers, operations, and spatial relationships is the hallmark of a mathematically developing mind (Figure 2.1).

FIGURE 2.1 ● Progression of Sophistication

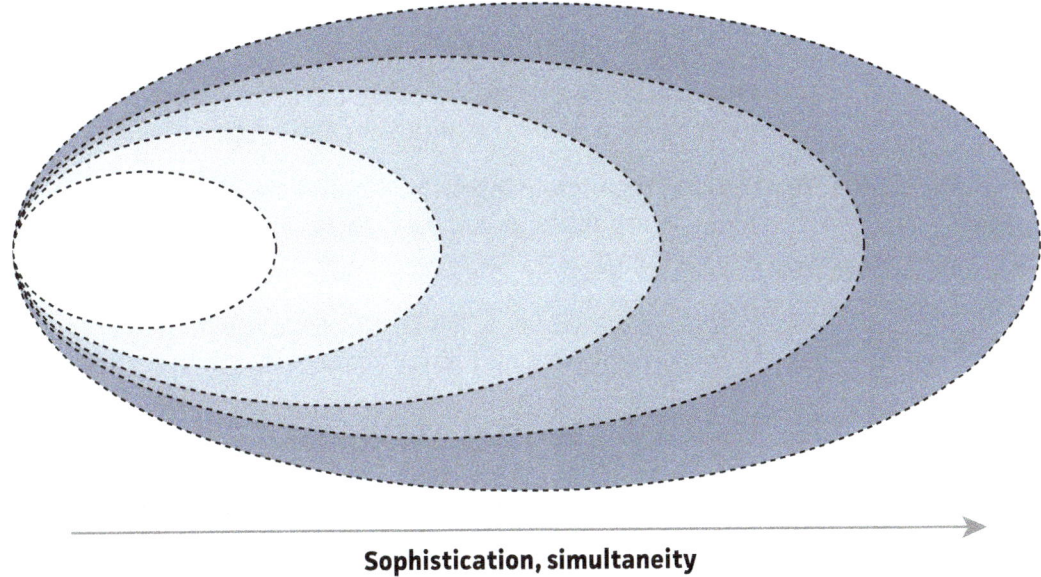

Sophistication, simultaneity

Source: Adapted from Math Is Figure-Out-Able at https://www.mathisfigureoutable.com/ with CC Attribution-NoDerivatives 4.0 International License.

In mathematics at its simplest—counting—a student may only be holding onto a single value of 1 at a time. This is counting three puppies, 1, 2, 3, and two puppies, 1, 2, and knowing there are five puppies, because you counted 1, 2, 3, 4, 5. At the opposite, more complex end of the K–12 spectrum is Functional Reasoning, in which a student holds onto, grapples with, and operates on entire sets of related numbers that vary in tandem. In between Counting Strategies and Functional Reasoning there are three other domains of mathematical reasoning.

Helping students develop from using beginning Counting Strategies to advanced Functional Reasoning is the essence of what it means to teach real math-ing. In a sense, much of what we do in the mathematics classroom should be in service of developing students' ability to hold on to and manipulate more and more data and ideas at the same time.

The ability to think and reason about multiple connected things simultaneously requires that students have experiences grappling with multiple things. This demands intentional and concerted effort to create classroom experiences to do just that. Students need time and space to consider and reject, reconsider and add in, revise and rework—all the while gaining clarity about the relationships involved.

TIP

When asking someone math questions, in order to determine how they are thinking about the question, keep your emphasis on their thinking, not the answer. Let the answer be a natural by-product of the discussion. Let them know that you're interested in how they think, the way they use the numbers or the structure in the problem.

As we discussed in the last chapter, algorithms are designed to minimize grappling. But that minimization severely limits essential growing experiences.

If this idea isn't quite clear yet, hold on. It will get clearer as we define each domain of reasoning in detail in this chapter.

To describe the increasing difficulty and complexity of thought involved as students develop more advanced reasoning, this book uses the word *sophisticated*.

- A domain of reasoning is more *sophisticated* (more complex, more complicated, and harder to develop) than the domain it is built from. The more sophisticated domain includes everything in the prior domains.

- A strategy is more *sophisticated* (more involved, interdependent) than another strategy if one must grapple with more things simultaneously or use more complex relationships than another strategy.

- A model is more *sophisticated* than another if it's more dense, representing more layers of connections and relationships.

As your students grow and develop a more mature, robust, multilayered, multifaceted web of interconnected relationships, the reasoning they are using is more advanced, more cognitively complex, more *sophisticated*. As we discussed in the preface, the use of the term *sophisticated* is in no way a value judgment of the people doing the reasoning.

I never describe people as more or less sophisticated.
It's about describing the reasoning that is less or more cognitively
complex as less or more sophisticated. Why don't I use the word complex?
Because complex *can have the connotation of elaborate, difficult,*
and complicated just to be complicated. I am instead
describing a developmental level and want to
convey the connotation of maturity.

Sophistication is also not a static way of being or a choice
we make all the time. Sometimes we weave in and out
of sophistication. We might reach for something less
sophisticated to get an answer while engaged in a bigger
problem. But other times we might reach for a more
sophisticated strategy to stretch ourselves. Part of math
class is helping students learn to choose judiciously.

WHY UNDERSTANDING THE HIERARCHY MATTERS

In basketball, there are a number of hierarchies of skills where it is pretty obvious which are prerequisite to others. Before you can shoot a layup, you need to be able to run and dribble at the same time. Before you can run and dribble, you need to stand and dribble. Before you can dribble at all, you need to be able to hold the ball correctly (Figure 2.2).

FIGURE 2.2 ● A Hierarchy of Learning to Play Basketball

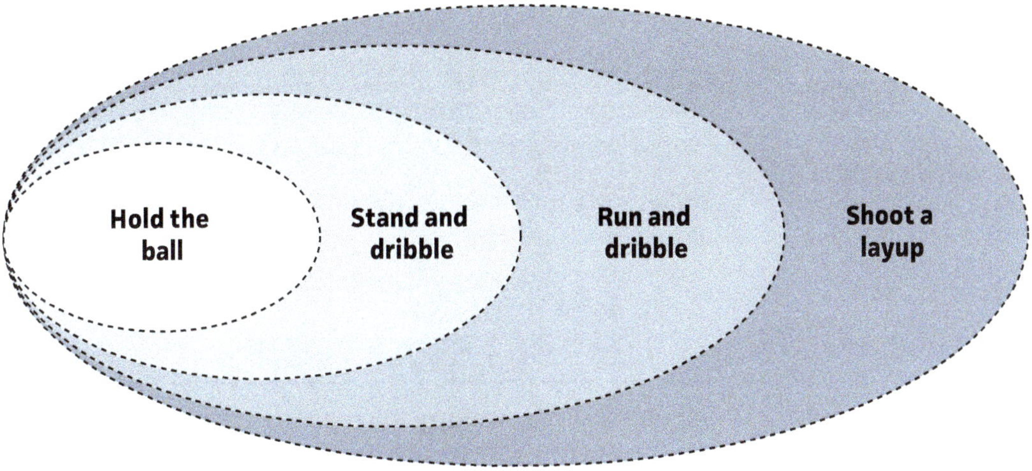

Hold the ball

Stand and dribble

Run and dribble

Shoot a layup

Similar hierarchies exist in mathematical reasoning, but they are often unknown. To extend the metaphor, we teach students how to hold the ball, ignore dribbling in all its forms, and expect students to successfully play a basketball game. We teach students to count, then expect them to do algebra without any ability to reason multiplicatively or proportionally. Predictably, we end up with whole generations of students who fear and hate math.

Some students figure it out on their own (these students are frequently under the second distortion, as described in the preface). Some have the natural talent or gumption to learn dribbling and shooting outside the classroom curriculum. But even they would reach far higher faster if the teaching helped them succeed instead of presenting a distortion to squint through.

The hierarchy of mathematical reasoning may not be as intuitive as basketball's (at least for me). This is true especially if you've been conditioned to think math is a hierarchy of algorithms to memorize rather than thresholds of understanding opening new horizons. It might be obvious that addition comes before multiplication, but why building Additive Reasoning helps then build Multiplicative Reasoning is less obvious, and how to teach it even less so.

Consider that it is perfectly possible to solve multiplication problems using an algorithm with nothing more sophisticated than counting by 1s. It will take a long time, but it is possible; no Multiplicative or even Additive Reasoning involved.

Solving a problem like 5 × 7 using a Counting Strategy could look like counting/drawing 5 circles, counting/putting 7 objects in each circle, then counting each object.

For those who make it past counting, 90% get stuck reasoning at best multiplicatively, never making it to Proportional or Functional Reasoning (Lamon, 2020).

So, let's dig into it. Let's lay out this hierarchy at the root of teaching real math-ing.

THE BEDROCK: COUNTING STRATEGIES

When students are tiny and solving tiny problems, they use Counting Strategies. Counting Strategies involve counting by 1, but using counting *strategies* is more than just being able to count or saying the counting sequence. It's about solving problems and, while solving, considering the numbers involved as sets of 1s (Figure 2.3).

FIGURE 2.3

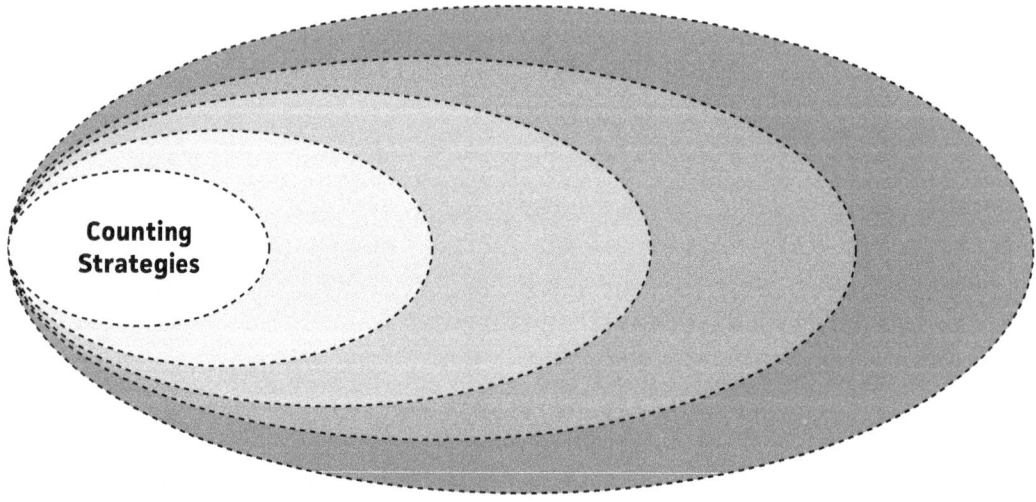

Counting Strategies

Source: Adapted from Math Is Figure-Out-Able at https://www.mathisfigureoutable.com/ with CC Attribution-NoDerivatives 4.0 International License.

A student might solve a problem like "I've got 4 books. My teacher gave me 3 books. How many books do I have?" with either of two Counting Strategies:

Counting 3 Times—Count out 4 objects, count out 3 objects, and count all the objects, 7.

Counting On—Start with 4. Count by 1s: 4. 5, 6, 7.

Either way, the student is counting one by one and using that one-by-one counting to solve a problem (Carpenter et al., 2014; Fosnot & Dolk, 2001a).

TRY IT

Ask young students questions like the following to determine if they are Counting 3 Times or Counting On. "Six butterflies are flying in the garden; 5 butterflies join them. How many butterflies are flying in the garden?" or "Nine balloons are floating, but 3 get popped. How many balloons are still floating?"

Counting 3 Times and Counting On are examples of less and more sophisticated reasoning within a single domain.

Counting On is more sophisticated than Counting 3 Times for several reasons. The student must be able to conceive of and hold the number 4 (the starting number of books), having an idea of four-ness. Then, as the student counts on 1 for each of the new books, the student must simultaneously keep track of when to stop. There is that simultaneity again.

As brilliant as Counting On is, we do not want to limit students to solving problems using Counting Strategies. For them to progressively consider more sophisticated mathematics, students need to learn to reason additively.

TIP

To determine if a child is Counting 3 Times or Counting On, ask for and listen to their reasoning. They might use fingers, objects, tally marks—watch to see how they use them. If they create the first number in some way, rather than starting from that number, they are probably Counting 3 Times.

TRY IT

Find 27 + 6 using the Counting On strategy. What do you have to keep track of as you solve? What if the problem is 27 + 16? How much are you keeping track of simultaneously?

THE FOUNDATION: ADDITIVE REASONING

Additive Reasoning is characterized by thinking in bigger jumps of numbers than one at a time. An additive reasoner considers numbers simultaneously as sets of 1s and combinations of other sets of numbers. It's about composing and decomposing numbers in additive chunks (Figure 2.4).

FIGURE 2.4 ● The Second Level of Sophistication in Mathematical Reasoning

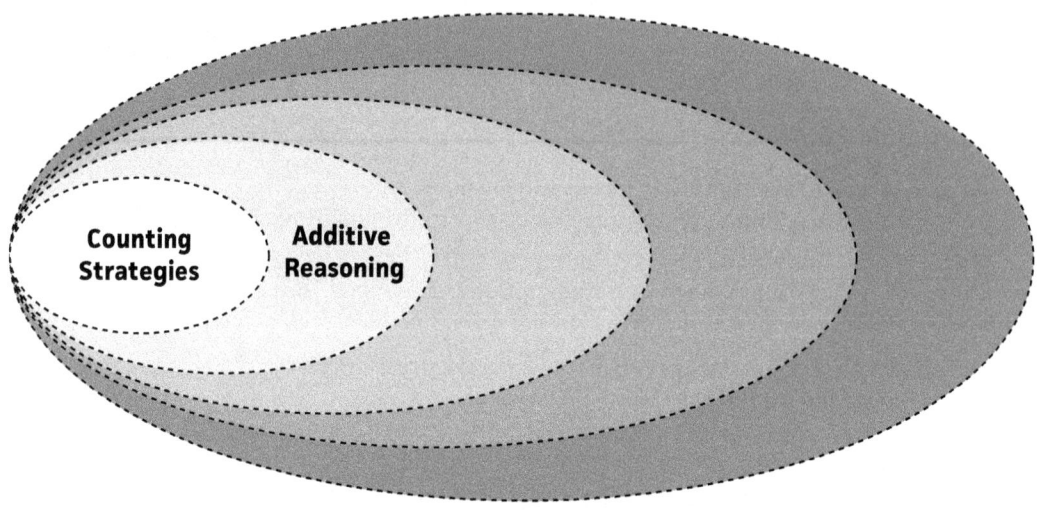

Source: Adapted from Math Is Figure-Out-Able at https://www.mathisfigureoutable.com/ with CC Attribution-NoDerivatives 4.0 International License.

Table 2.1 represents the difference between using Counting Strategies and Additive Reasoning, with the Additive Reasoning shown on two different models.

TABLE 2.1 ● What's the Difference Between Counting Strategies and Additive Reasoning?

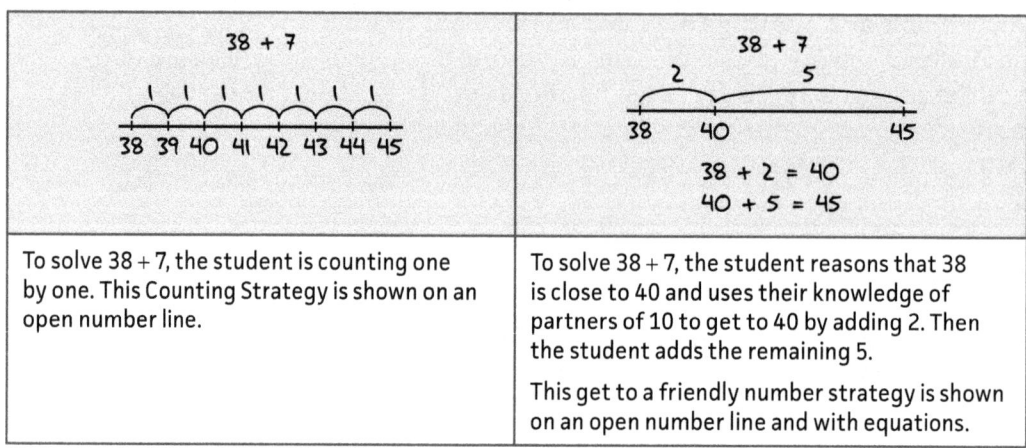

To solve 38 + 7, the student is counting one by one. This Counting Strategy is shown on an open number line.	To solve 38 + 7, the student reasons that 38 is close to 40 and uses their knowledge of partners of 10 to get to 40 by adding 2. Then the student adds the remaining 5.
	This get to a friendly number strategy is shown on an open number line and with equations.

The strategy shown for 38 + 7 in the table can be called getting to a friendly number *and is one of the major addition strategies to develop.*

Additive Reasoning for single-digit numbers is about using relationships you already own to find ones you do not yet know. For example, to find 7 + 8, students could reason additively as follows:

- I know 7 + 7 is 14, so 7 + 8 must be 1 more, 15.
- Since 8 + 8 is 16, 7 + 8 is 1 less, so 15.
- I know 7 + 3 = 10, so 5 more is 15.
- 7 + 10 is 17, so 7 + 8 must be 2 less, so 15.

The strategies shown for 7 + 8 represent the three major single-digit additive strategies to help students to develop: using doubles, get to 10, *and* add 10 and adjust.

FREQUENTLY ASKED QUESTIONS

Q: Many of my students can find their single-digit addition facts really quickly counting by ones. It can take longer for them to use an additive strategy. Does it matter?

A: Yes! It matters. It's a great starting point for those students to solve single-digit addition problems counting by ones, but it cannot be where we leave them. We need to help students develop the additive relationships necessary to use Additive Reasoning. Remember, it's not about getting answers; it's about building stronger brains that are capable of more complex thinking. Give students many experiences using the additive strategies so that if they do not just know a fact, they use an additive relationship to figure it out. Make it clear that if a student does not just know a fact, their job is to figure it out using an additive strategy.

TRY IT

Ask several young students 7 + 8 to determine if they are using a Counting Strategy or Additive Reasoning. Watch to see if they count by ones (Counting Strategy) or use larger jumps of numbers, building off an addition fact they already know (Additive Reasoning).

We dive into Additive Reasoning and how teaching to algorithms undermines it in more depth in Chapter 3.

THE LAST STOP FOR MOST: MULTIPLICATIVE REASONING

Multiplicative Reasoning is more than just rote-memorizing multiplication facts or getting an answer to a multiplication or division problem. It is also more mentally sophisticated than Additive Reasoning (Figure 2.5).

FIGURE 2.5 ● The Third Level of Sophistication in Mathematical Reasoning

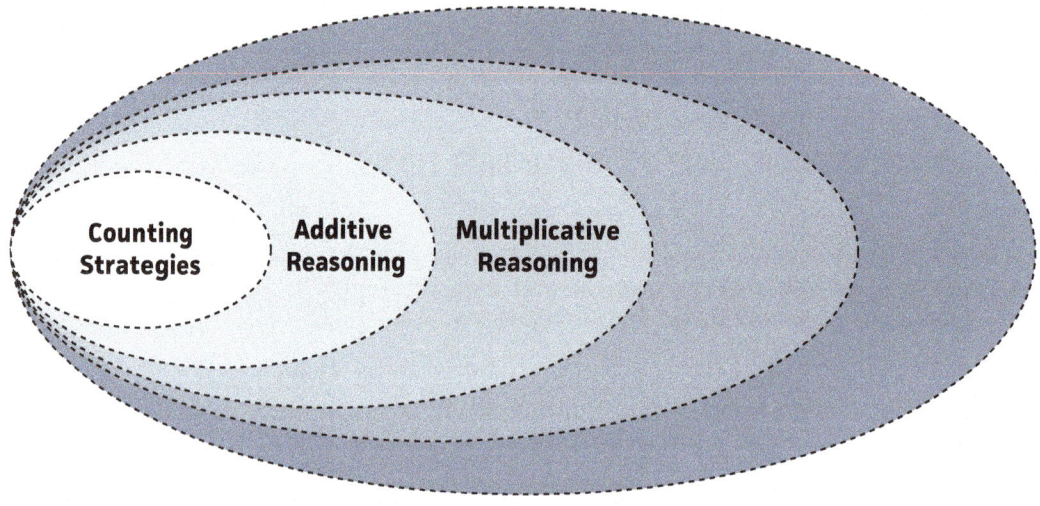

Source: Adapted from Math Is Figure-Out-Able at https://www.mathisfigureoutable.com/ with CC Attribution-NoDerivatives 4.0 International License.

Multiplicative Reasoning starts with thinking about bigger chunks of the product than one group at a time. Table 2.2 represents the difference between using Additive Reasoning and Multiplicative Reasoning.

TABLE 2.2 ● What's the Difference Between Additive Reasoning and Multiplicative Reasoning?

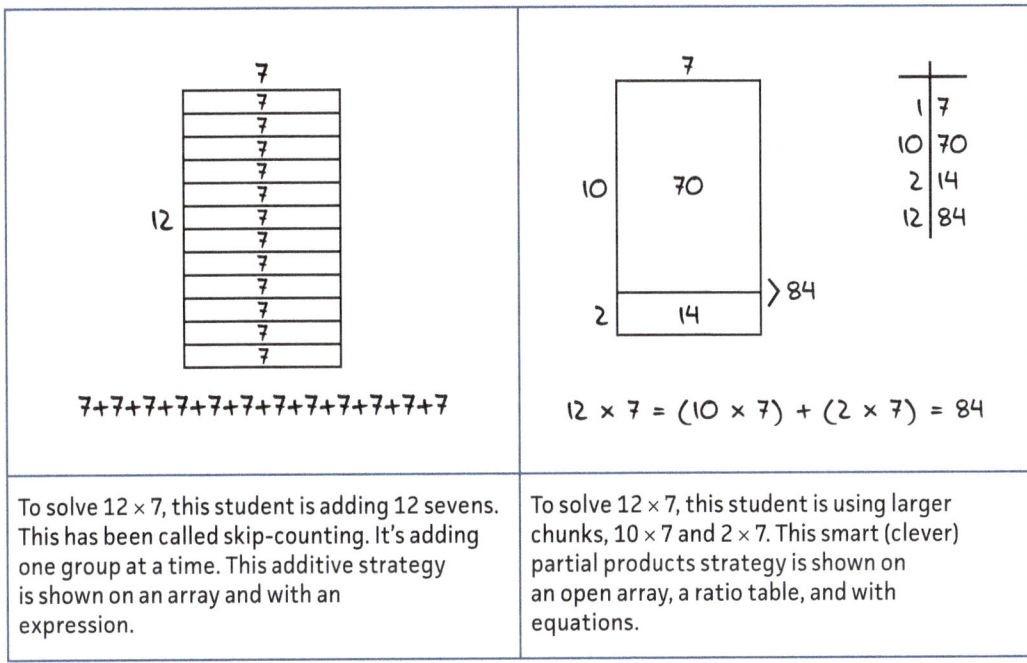

7+7+7+7+7+7+7+7+7+7+7+7	$12 \times 7 = (10 \times 7) + (2 \times 7) = 84$
To solve 12 × 7, this student is adding 12 sevens. This has been called skip-counting. It's adding one group at a time. This additive strategy is shown on an array and with an expression.	To solve 12 × 7, this student is using larger chunks, 10 × 7 and 2 × 7. This smart (clever) partial products strategy is shown on an open array, a ratio table, and with equations.

The strategy shown for 12 × 7 in Figure 2.5 can be called
smart (clever) partial products *and is one of the*
major multiplication strategies to develop.

If Counting Strategies are all about counting one by one, and Additive Reasoning is thinking about bigger jumps than one at a time, then Multiplicative Reasoning is thinking about bigger chunks than one group at a time. It's about grouping the groups simultaneously (Carpenter et al., 2014; Fosnot & Dolk, 2001b). There's that increasing simultaneity happening.

These relationships can be represented by arrays, as shown in Figure 2.6.

FIGURE 2.6 ● Arrays

Counting	Additive	Multiplicative

1 at a time	1 group at a time	Grouping the groups

FREQUENTLY ASKED QUESTIONS

Q: Many of my students can find their single-digit multiplication facts quickly by skip-counting. It can take longer for them to use a multiplicative strategy. Does it matter?

A: Yes! It matters. It's a great starting point for those students to solve single-digit multiplication problems by skip-counting, but it cannot be where we leave them. We need to help students develop the multiplicative relationships necessary to use Multiplicative Reasoning. Remember, it's not about getting answers; it's about building stronger brains that are capable of more complex thinking. Don't let students get trapped in skip-counting. Give students many experiences using the multiplicative strategies so that they do not just know a fact; they use a multiplicative relationship to figure it out. Make it clear that if a student does not just know a fact, their job is to figure it out using a multiplicative strategy.

Q: What about the load on working memory? Shouldn't we just have students rote-memorize the facts so they can use them in bigger problems?

A: The short answer is no. The purpose of math class is to build mathematical reasoning, and we can do that while reasoning about the single-digit facts. Students can use the same reasoning they build to find 9×7 as $10 \times 7 - 7$ to find 9×57 as $10 \times 57 - 57$, which not only gets them answers to two multiplication problems, but also builds place value, magnitude, rounding, estimation, 10 times anything, and Multiplicative Reasoning. We can take all the time that has been spent drilling students on the steps of the multiplication algorithm and spend it more fruitfully here. The longer answer is in Chapter 4. Keep reading!

An example of Multiplicative Reasoning with larger numbers is finding 45 × 18 by thinking about 20 × 45 but realizing that's too much, so removing the extra 2 × 45: 45 × 18 = 18 × 45 = 20 × 45 − 2 × 45 = 900 − 90 = 810.

Multiplicative Reasoning can also get more multiplicative. For example, one can factor and re-associate the factors to make an easier problem to solve:

45 × 18 = (9 × 5) × (2 × 9) = 9 × (5 × 2) × 9 = 9 × 9 × 10 = 810

The multiplication strategy of factoring the factors and reassociating the resulting factors can be called flexible factoring. It is a fantastic strategy for students to develop so they recognize factors as helpful in multiplicative situations.

For the vast majority of students who make it past Counting Strategies, Multiplicative Reasoning is their last stop. Without purposeful instruction, this is the most sophisticated reasoning most students will achieve on their own (Lamon, 2020).

We dive into Multiplicative Reasoning and how teaching to algorithms undermines it in more depth in Chapter 4.

BEYOND THE HORIZON: PROPORTIONAL REASONING

As many as 90% of adults in the United States are not reasoning proportionally (Lamon, 2020). But the vast majority of those same adults still took classes in algebra, geometry, and more (Figure 2.7).

TIP

When asking people math questions to determine how they are thinking about the problem, have paper and pen/pencil handy. Do not make people keep all their thinking in their head. Allow them to write to keep track of the relationships they are using. To paraphrase Cathy Fosnot, mathematical reasoning is not about doing it all in your head, it's about doing it *with* your head (Fosnot & Uittenbogaard, 2007). It's perfectly acceptable to keep track of your mental thinking.

FIGURE 2.7 ● The Penultimate Stage of Mathematical Reasoning

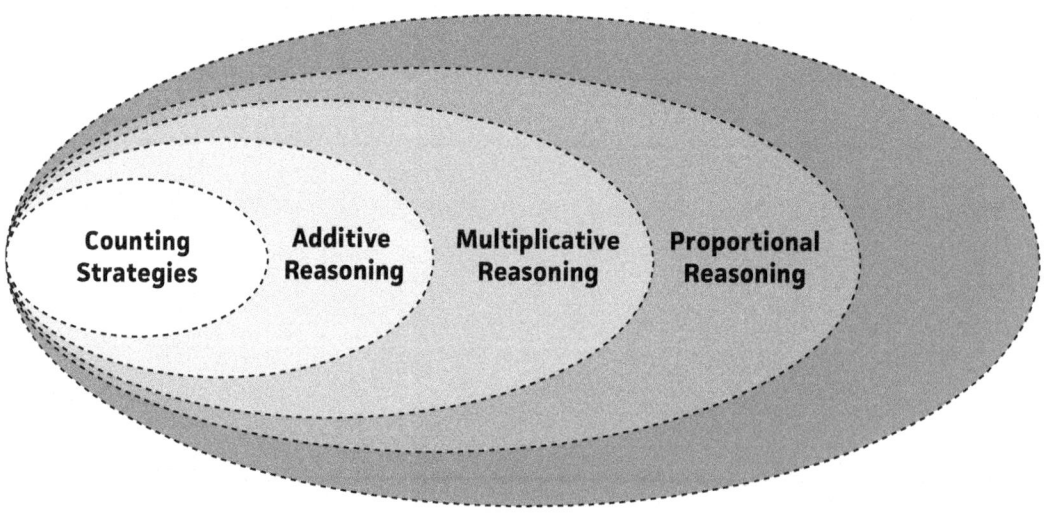

Counting Strategies · Additive Reasoning · Multiplicative Reasoning · Proportional Reasoning

Source: Adapted from Math Is Figure-Out-Able at https://www.mathisfigureoutable.com/ with CC Attribution-NoDerivatives 4.0 International License.

Is it any wonder math is so misunderstood, feared, or even hated? All of those subjects, which require Proportional Reasoning to understand, add up to years spent trying to guess what the textbook means and punishment for failing to do so.

Proportional Reasoning is more sophisticated than previous domains because, again, we are asked to consider more things simultaneously. Whereas in multiplication we think about grouping the groups—the number in a group and the number of groups at the same time—in Proportional Reasoning we consider the simultaneous relationship between two multiplicative quantities. Not only are we scaling, but we're also scaling in tandem.

This is the first domain where applying reasoning from a previous level incorrectly shows up in force. Because Proportional Reasoning is by nature multiplicative, if students have not constructed and developed Multiplicative Reasoning, they will attempt to use Additive Reasoning or Counting Strategies, both of which will not work. For example, if 4 slices of pizza is $5, but the customer wants to buy only 3 slices, a student reasoning additively might think, 4 slices minus 1 slice is 3 slices, so the corresponding cost would be $5 – $1, so $4.

This is why distorting mnemonic devices like "Ours is not to reason why, just invert and multiply" rear their ugly heads so often in middle school. Students left behind in Additive Reasoning or earlier simply aren't yet equipped to think about fractions, ratios, and proportions. Therefore, many teachers have tried to make the best of the situation and found ways to help students get answers without reasoning. However, we can do better.

We dive into Proportional Reasoning and how teaching to algorithms undermines it in more depth in Chapter 5.

THE ALIEN FRONTIER: FUNCTIONAL REASONING

The domain of Functional Reasoning is all about bivariate covariation—where two variables are varying in tandem. Whereas in Proportional Reasoning, two quantities are varying in tandem, in Functional Reasoning, two *variables* are varying in tandem (Figure 2.8). The simultaneity keeps compounding.

FIGURE 2.8 ● The Full Spectrum of Mathematical Reasoning

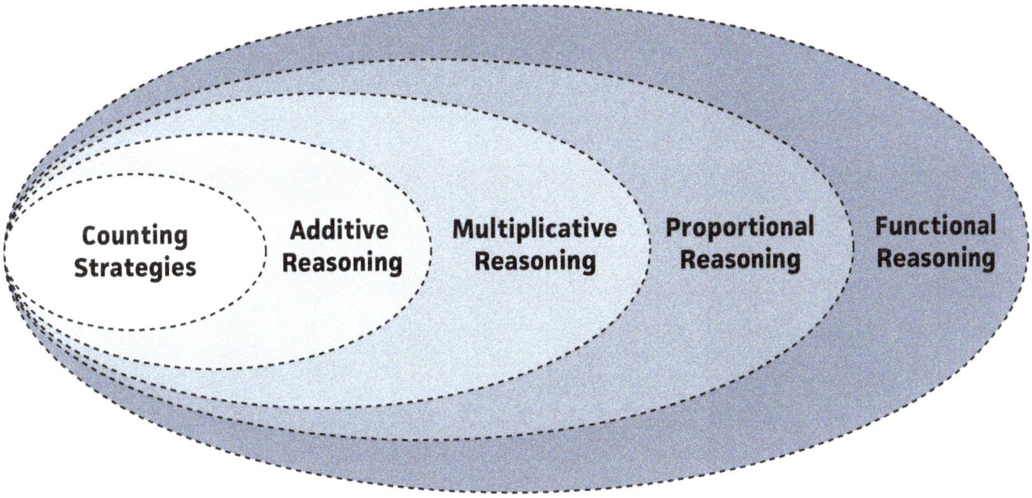

Counting Strategies

Additive Reasoning

Multiplicative Reasoning

Proportional Reasoning

Functional Reasoning

Source: Adapted from Math Is Figure-Out-Able at https://www.mathisfigureoutable.com/ with CC Attribution-NoDerivatives 4.0 International License.

At this point, if a student has not been building their simultaneity skills, asking them to work with functions is a lot like asking me (who can't juggle at all) to juggle four flaming knives all at once, except it's not four knives, it's $4x$ knives, where 4 is the instantaneous velocity of all the knives at time x, and I throw the knives from 3 in. above the ground to 7 in. above the ground before catching them.

For many students, functions are that opaque, and make about as much sense. Saying we are making things easier by assuming no air resistance makes no difference, because it's already so much more than most students can handle they weren't even thinking about including air resistance in the first place.

We dive into Functional Reasoning and how teaching to algorithms undermines it in more depth in Chapter 6.

PUTTING IT ALL TOGETHER: THE DEVELOPMENT OF MATHEMATICAL REASONING GRAPHIC

Now that we've introduced the parts, let's discuss how they fit together. As you can see in Figure 2.9, each of the domains of reasoning we have discussed are related to each other in that they subsume earlier stages.

FIGURE 2.9 ● The Full Spectrum of Mathematical Reasoning

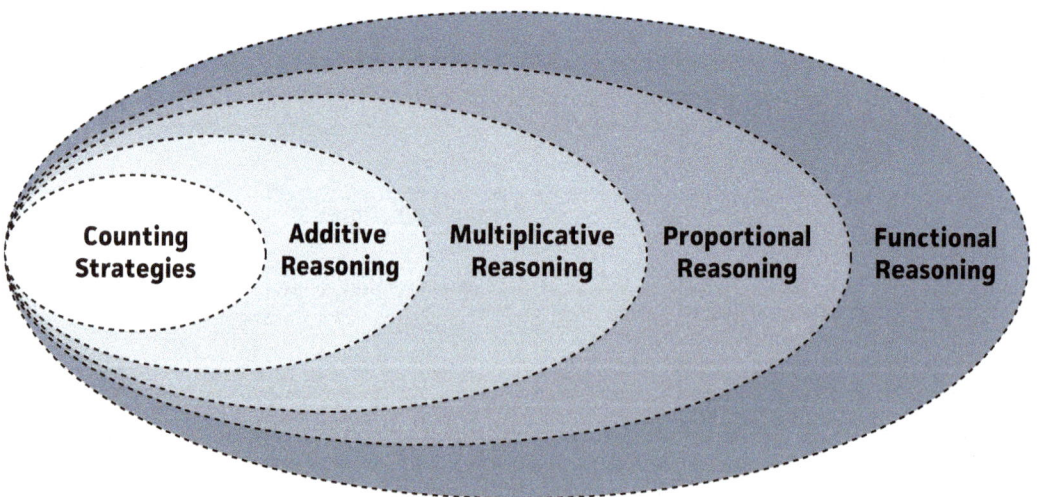

Source: Adapted from Math Is Figure-Out-Able at https://www.mathisfigureoutable.com/ with CC Attribution-NoDerivatives 4.0 International License.

I've chosen to depict the domains of reasoning as concentric ovals to represent their hierarchical nature—they are based on and build from each other. Proportional Reasoning is foundational to Functional Reasoning, and so forth, but

I've also separated each domain with a dotted line to suggest the dividing line between each domain is not solid, nor easily demarcated. These are boundary areas where one domain bleeds into another. Students will go in and out of less and more sophisticated reasoning until they have built the next domain solidly.

For example, to solve 6×7, a student might start by skip-counting (Additive Reasoning) to find the answer to the more manageable 3×7. That would be 7, 14, 21. Having found 3×7 with Additive Reasoning, they might then find 6×7 by doubling (early Multiplicative Reasoning) the now established 3×7 to get 6×7.

$3 \times 7 = 21$, therefore $6 \times 7 = 42$.

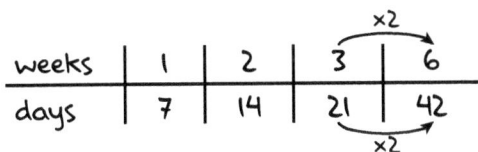

That same student might reason multiplicatively about a different problem, 7×8.

Seven 8s is five 8s plus two more 8s.

7×8 is five 8s plus two more 8s.

$7 \times 8 = 5 \times 8 + 2 \times 8$

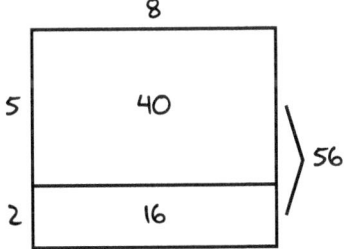

But as numbers get larger and more unwieldy, that same student might skip-count, additively reasoning, to find 6×25.

25, 50, 75, 100, 125, 150. So, 150.

The numbers get bigger and the student reverts to what they are most familiar with.

A less experienced student might solve 7 + 5 as 7 + 3 and add 2 more. But ask that same student to find 27 + 5 and they might count by 1s: 27, 28, 29, 30, 31, 32, despite both addition problems using the same partners of 10. The numbers get bigger and the student reverts to the less sophisticated counting by 1s, a Counting Strategy.

REASONING DOESN'T EXPIRE

The domains are arranged left to right because it is critical to understand that one domain is not inferior to another, merely less sophisticated. This is not a ladder to climb; it is a toolbox with prerequisites. A Functional Reasoning pro will still encounter situations where they need to count or use Additive or Multiplicative Reasoning. If you're counting puppies, there is no reason to generalize the number of puppies in the room with a function. At the same time, someone who cannot count the puppies could not generalize the number of puppies with a function if the situation did call for that level of sophistication.

Furthermore, each domain contains its own hierarchy of less to more sophisticated forms of that reasoning. Not all Multiplicative Reasoning is as relatively sophisticated as all other Multiplicative Reasoning. Exponents require more sophistication than doubling. See Figure 2.10 for a further breakdown, but note that these are not a linear path. They overlap and intersect, and students can travel different paths.

FIGURE 2.10 ● Breakdown of Multiplicative Reasoning Domain

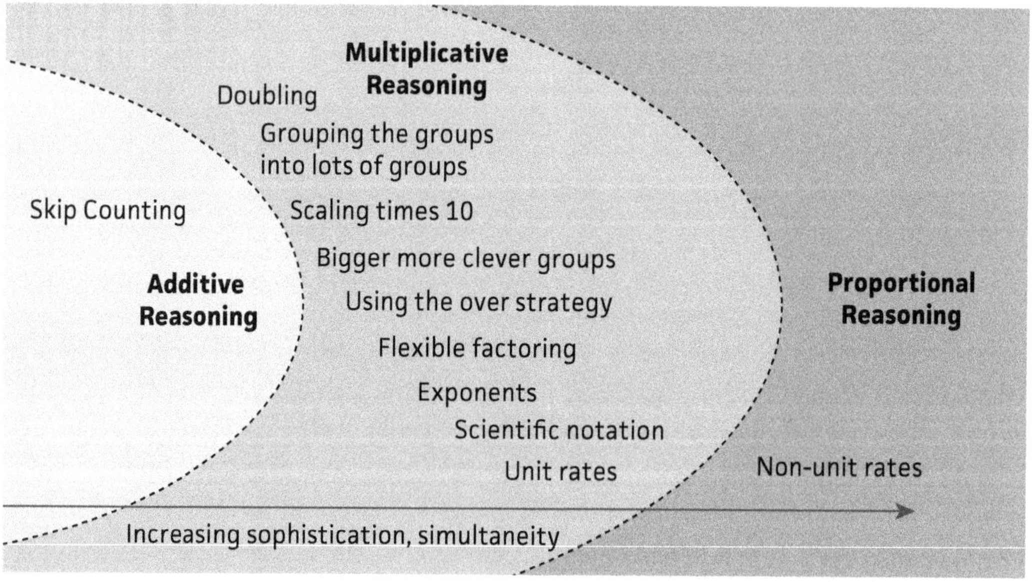

Source: Adapted from Math Is Figure-Out-Able at https://www.mathisfigureoutable.com/ with CC Attribution-NoDerivatives 4.0 International License.

Understanding this breakdown is helpful for identifying how each of your students is reasoning and for sequencing lessons and topics. Skipping one building block can be done, but doing so requires a thorough understanding of how the missing pieces will influence student thinking.

Similarly, learning only the rudiments of Multiplicative Reasoning or any other domain is not sufficient (again, see Figure 2.10). The basics are essential but not enough for success, whether defined by standards or understanding.

Remember the doctor whose son had the rudiments of multiplicative thinking? As the doctor-mom correctly identified, the son was adding to find multiplication facts. His unsophisticated reasoning is a great start but not enough. This is where he needs a teacher to know the hierarchy and to help him develop more and more sophisticated Multiplicative Reasoning.

SPATIAL, ALGEBRAIC, AND STATISTICAL REASONING

In addition to the previously discussed five hierarchical domains in DMR, there are three longitudinal domains: Spatial, Algebraic, and Statistical, as shown in Figure 2.11.

FIGURE 2.11 ● The Full Spectrum of Mathematical Reasoning With Longitudinal Domains

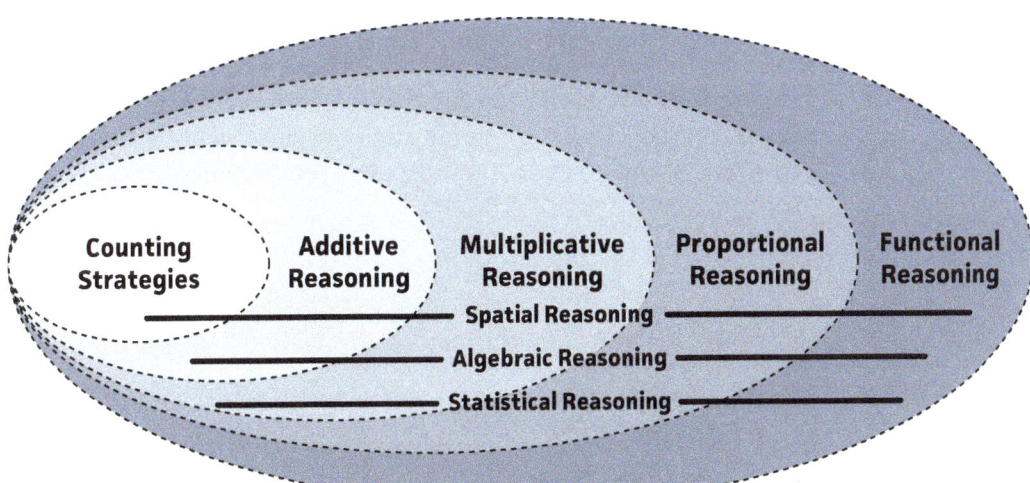

Source: Adapted from Math Is Figure-Out-Able at https://www.mathisfigureoutable.com/ with CC Attribution-NoDerivatives 4.0 International License.

These are represented as horizontal lines crossing the other domains because, unlike the others, they can and should be built concurrently with the other domains and do not necessarily have prerequisites. They benefit from and help build the five hierarchical domains.

For example, the use of visual models like open number lines and open arrays (area models) work well to simultaneously develop Spatial Reasoning along with the other domains.

When students see the number 64, we want to help them develop many notions about 64—not just the digits 6 and 4. We can put the number on an open number line, a model, with many related important numbers around it. We can represent it as the area of several rectangles.

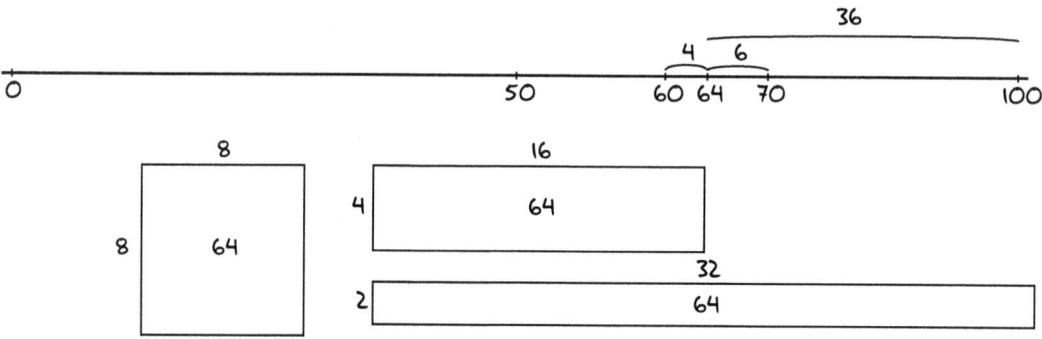

Using open number lines to represent student thinking about addition and subtraction strengthens students' sense of place value, neighborhood and nearness, rounding, etc.

Figure 2.12 shows a student thinking about 337 + 95 by adding too much, 100, then adjusting back 5. This shows students thinking about numerical and spatial relationships, place value, rounding, and friendly numbers.

FIGURE 2.12 ● Adding Too Much and Adjusting Back

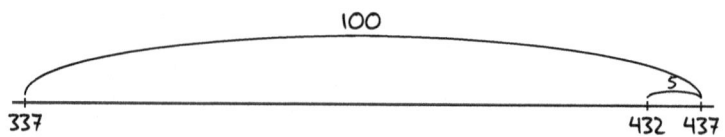

Using open arrays (area models) to represent student multiplication and division strategies strengthens students' sense of dimension, measurement, breadth, and depth.

This shows thinking about 33 × 98 by multiplying by too much, 100, then adjusting back two groups of 33. Students are thinking about numerical and 2D spatial relationships, groups of groups, place value, rounding, and friendly numbers.

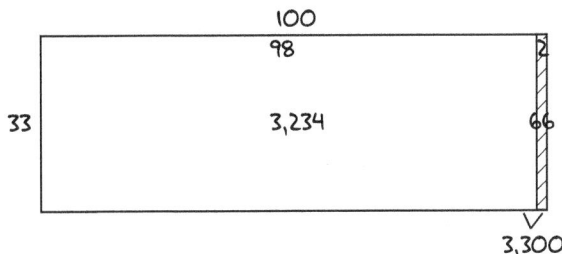

Spatial Reasoning is all about the visual, geometric relationships of shapes and dimensions, measurement and location, graphs and trends. We will use spatial models to make visible much of the mathematics in the rest of the book.

Algebraic Reasoning is all about generalizing and working with those generalizations. For younger learners, this can look like reasoning that using 10 is helpful for 9s, as in 8 + 10 is 18 so 8 + 9 is one less, 17. For older learners, this can look like generalizing that if you take from one addend to give to the other to make a friendly number, the sum stays the same. For older learners this looks like $(a + c) + (b - c) = a + b + (c - c) = a + b + 0 = a + b$. Every time we generalize we are using algebraic reasoning, from deciding which strategy is a good fit for a problem to evaluating expressions to solving equations.

Statistical Reasoning is all about data, inferences, and predictions. Many now refer to data science, especially in the wake of big data with things like large language models and AI. Statistical reasoning is one of the most critical skills in our modern age. Consider that many, if not all, doctors must take calculus but, unfortunately, never take statistics. Yet they need to use statistical reasoning often in their medical practice to do everything from reading charts and graphs to weigh risks and using statistics to determine which medicines to prescribe. While statistical reasoning is as important to build as Spatial and Algebraic Reasoning, the scope of that conversation is out of the bounds of this book.

To maximize effectiveness, mathematics teaching must take advantage of the reciprocal reinforcement and development created by building both these longitudinal (Spatial, Algebraic, and Statistical) and hierarchical (counting through functional) reasonings simultaneously. This synergy has already been at work and will be at work with the various examples and models used in the book.

HOW TO READ THE REST OF THIS BOOK

As you read the following chapters, you will find sections that go into detail about the traps of a few major algorithms. You might be tempted to skip these parts, but use them as training to spot similar traps in other step-by-step procedures. Learning to spot the traps helps us as teachers seek ways to better understand and to help our students better understand.

Discussion Questions

1. What is the difference between the hierarchical domains (Counting, Additive, Multiplicative, Proportional, Functional) and the more longitudinal domains (Spatial, Algebraic, Statistical)?

2. Is there a place in the hierarchy when your own reasoning fell off in school? Can you identify why?

3. With which sophistication/complexity might you be reasoning now? Why do you think so?

4. With which sophistication/complexity do you think the majority of your students are reasoning? Why do you think so?

5. How might understanding this hierarchical progression help you in your teaching?

TRY IT IN YOUR CLASSROOM

How Are Your Students Reasoning?

Purpose:

To determine how your students are reasoning, ask students these questions. Listen carefully to the way they are thinking about the problem, how they are using relationships. It's not about whether the answer is correct. It's about how they reason to find it.

Routine:

What is 8 + 5? (For younger students)

Counting Strategies: Counting by 1s

- Count out 8 objects (tallies, etc.), count out 5. Put together, count the whole set.

- Start with 8 and count 9, 10, 11, 12, 13. Counting On.

What is 58 + 5?

Counting Strategies: Counting by 1s

- Count out 58 objects (tallies, etc.), count out 5. Put together, count the whole set.

- Start with 58 and count 59, 60, 61, 62, 63. Counting On.

Additive Thinking: Using jumps bigger than 1s

- From 58, add 2 to get to 60. Add the remaining 3 to get 63.

(Continued)

(Continued)

What is 16×9?

Counting Strategies: Counting by 1s

- Count out 16 groups of 9 objects (tallies, etc.) or 9 groups of 16 objects, one at a time. Put together, count the whole set.

Additive Thinking: Adding one group at a time

- Skip-count by 16s or 9s (16, 32, 48, ... 144 or 9, 18, 27, ... 144)

Multiplicative Reasoning: using bigger chunks than one group at a time

- Think about 16 9s as 10 9s and six more 9s.
$10 \times 9 + 6 \times 9 = 90 + 54 = 144$

- Think about nine 16s as 10 9s subtract one 16.
$10 \times 16 - 1 \times 16 = 160 - 16 = 144$

- Think about equivalent problems by doubling/halving.
$16 \times 9 = 8 \times 18 = 4 \times 36 = 2 \times 72 = 144$

Solve for x: $\left(\frac{5.5}{2.2}\right) = \left(\frac{1.25}{x}\right)$

Multiplicative Reasoning:

- Find $2.2 \times 1.25 \div 5.5$. If done with the traditional algorithms for multiplication and division, the Multiplicative Reasoning can deal at best with single-digits.

Proportional Reasoning:

- Scale in tandem to find helpful ratios.
$\left(\frac{5.5}{2.2}\right) = \left(\frac{55}{22}\right) = \left(\frac{5}{2}\right) = \left(\frac{1.25}{0.5}\right)$, so $x = 0.5$

CHAPTER 3

The Trap of Addition and Subtraction Algorithms

FIGURE 3.1 ● The first domains of reasoning are Counting Strategies and Additive Reasoning

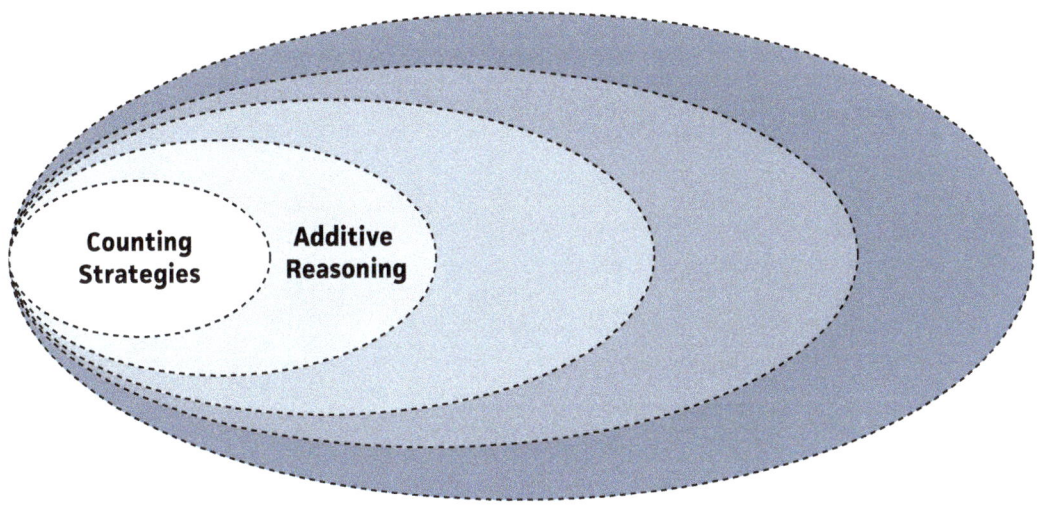

Source: Adapted from Math Is Figure-Out-Able at https://www.mathisfigureoutable.com/ with CC Attribution-NoDerivatives 4.0 International License.

W ait," says Holly to another graduate student. "Are you saying that when you see problems like 46 + 99, you don't think about lining up the numbers and then see digits and dots? I do . . ." she trails off. She sits up abruptly. "Is that why I've struggled in math? Because I was left in digits and dots when I could have been thinking the way I just did in that String?"

Let me back up.

It's 2015, and I'm teaching graduate students at Texas State University on the differences between Counting Strategies and Additive Reasoning. Touch Math comes up as a pertinent example.

Touch Math is a well-intentioned procedure/process/trick that appears in special education and early grade math classes, especially when students are struggling. With Touch Math, students often get the answers to addition/subtraction problems but build none of the reasoning they will need to move forward.

Touch Math

1 2 3 4 5

Each numeral has a set of dots for that number.

Memorize where the dots are. When counting, touch each dot.

Students are instructed to write the numerals with dots for the digits in the problem and touch the dots while counting each one to get the answer. For example, 3 + 5 looks like writing the numeral 3 with three dots and the numeral 5 with five dots, and then counting the dots one by one: 1 → 2 → 3 (move to the 5 and keep counting) → 4 → 5 → 6 → 7 → 8. The answer is 8.

To add 3 + 5:

Touch each dot on the 3 as you say 1, 2, 3. Continue touching the dots on the 5 as you say 4, 5, 6, 7, 8.

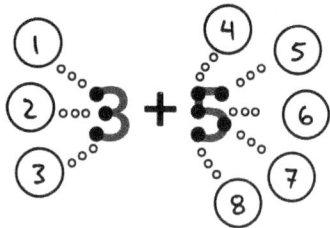

The class correctly identifies that this method requires students to count one by one, even for multidigit problems. Use of columns facilitates getting answers to multidigit problems but obscures place value in the process. Even then, there are so many dots. So. Many.

Like any other method that relies on memorized steps, depending on memorized patterns of dots and counting one by one, each and every time, becomes a trap. Students are too frequently left with atrophying reasoning muscles when they need to be bulking up those muscles.

Holly was one such student.

In this graduate class, I facilitate a Problem String where students are encouraged to build/use Additive Reasoning. One set of the problems is 46 + 100 followed by 46 + 99. We discuss that Additive Reasoning is not counting one by one, like the dots; rather, it is thinking about bigger jumps of numbers—holding and dealing with groups of numbers.

46 + 100 = 146

46 + 99 = 145

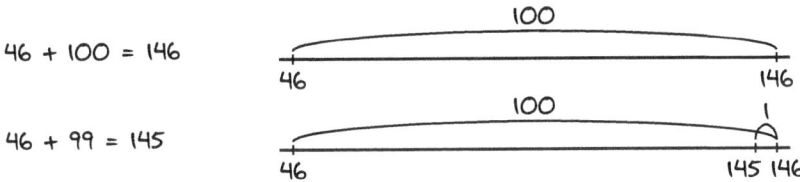

Eyes open to what could have been. Holly is upset, and understandably so. Not that she ascribes malice to any of her teachers, but malice is not required for very real, very negative consequences to occur. Trapped by an algorithm, Holly had been left to count her way through her math classes when she could have been reasoning.

Holly has always been a fantastic teacher. Now, she is gaining the tools to make sure none of her students will be trapped like she had been. Her fourth-grade classroom has become a powerhouse of Additive and Multiplicative Reasoning.

HOW ARE YOU THINKING ABOUT ADDITION RIGHT NOW?

Did you learn something like Touch Math? Was your addition instruction primarily centered around algorithms? How do you think about addition problems now?

In this section, focus on the way your brain handles these problems. Solve the problem, then read the descriptions underneath and choose one that best fits your thinking.

How do *you* think about 58 + 5 and 87 + 98?

58 + 5

- If you started at 58 and counted by ones: 59 → 60 → 61 → 62 → 63 (not picturing or writing numbers with arrows, but counting one by one, whether in your head or somewhere else), then you were using a Counting Strategy.

- If you thought about getting from 58 to 60 and that takes 2, then 3 more to 63, then you were using an additive strategy.

87 + 98

- You lined up the digits in columns and thought: 7 + 8. 7. 8 → 9 → 10 → 11 → 12 → 13 → 14 → 15. Write down the 5, carry the 1. Now, 8 + 9 + 1: 8. 9 → 10 → 11 → 12 → 13 → 14 → 15 → 16 → 17 → 18. Write down the 18 next to the 5, 185. If you did these steps and found the single-digit additions by counting by ones, then you were using a Counting Strategy.

- You lined up the digits in columns and thought: 7 + 8 is like 7 + 10, back up 2, so 15. Write down the 5, carry the 1. Now, 8 + 9 + 1. And 9 + 9 is 18, so 8 + 9 is 17, add the carried 1 is 18. Write down the 18 next to the 5, 185. If you did these steps and reasoned about the single-digit additions, you were using Additive Reasoning with single-digit numbers.

TIP

Use these orienteering/ navigating questions with students to determine how they are reasoning.

- If you thought 87 + 98 is like 87 + 100 = 187, but it's 2 too much, so 185, you were using Additive Reasoning with multidigit numbers.

WHAT ARE COUNTING STRATEGIES?

If someone is counting by 1, one at a time, they are using a Counting Strategy. This strategy can be used to solve many kinds of problems, especially if employed inside an algorithm, which means students can use counting to get correct answers (that they don't understand) to problems (that they don't understand).

In Table 3.1, you see two common, result-unknown problem types and corresponding Counting Strategies to solve them (Carpenter et al., 2014). Notice how each Counting Strategy involves one-by-one counting.

TABLE. 3.1 ● Counting Strategies for Combining and Separating Problems

	COUNTING STRATEGIES	
THE PROBLEM	**COUNTING 3 TIMES** (Less Sophisticated)	**COUNTING ON, OR BACK** (More Sophisticated)
You have 6 balloons, then find 3 more. How many do you have now? (Combining—addition)	Count 6 objects, count 3 objects, combine them, then count 9 objects.	Start with 6. Count On by ones: 6. 7 → 8 → 9.
You have 7 balloons, then pop 3. How many are left? (Separating—subtraction)	Count 7 objects, count 3 of those objects out, then count the remaining objects left.	Start with 7. Count Back by ones: 7. 6 → 5 → 4 → so 4. Or, 7 → 6 → 5, so 4.

Counting 3 Times, whether counting on fingers, putting out counters, or drawing dots, leaves you counting 3 times, every time. Students will get the correct answers, but in a tedious manner, without strengthening their brains. Counting On or Counting Back is more sophisticated than Counting 3 Times because students must conceptualize the starting number enough to start with it, rather than counting up to it. Then they must do a few things simultaneously: count, keep track of the count, and keep track of when to stop.

To Count On

7 + 4

1. Conceptualize 7.
2. Say the counting sequence while keeping the goal of adding 4 in mind.

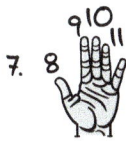

7. 8

3. Stop because 4 fingers are up.

In this example of 7 + 4, a student must simultaneously keep track of the counting sequence and how many they have counted: "7, 8 and 1, 9 and 2, 10 and 3, 11 and 4. Stop because I hit 4, so the answer is 11."

We don't want students to be limited to this Counting On or Back, but at least they have stretched and strengthened their brains to deal with things simultaneously compared to Counting 3 Times.

You might find it interesting that Counting Strategies can be used for more than addition and subtraction problems. In Table 3.2, notice that students can solve multiplication and division problems using Counting Strategies. The type of reasoning, Counting Strategies in this case, is not defined by the problem type being solved. It's about the way a student is dealing with the numbers in the problems: in this case, one by one by one.

TIP

How can you tell if a student is Counting 3 Times or Counting On/Back? If a student counts out both addends, creating them both in some way, they are Counting 3 Times. If the student says one of the addends and then counts from there, they are Counting On. If the student says the total and counts back from there, they are Counting Back.

TABLE 3.2 ● Counting Strategies for Multiplication and Division Problems

PROBLEM TYPE	THE PROBLEM	COUNTING STRATEGY
Multiplication	4 people, each get 2 balloons. How many total balloons needed?	Count 4 to represent the 4 people. Count 2 in each of the 4 places. Count all of the objects one by one.
Division	We have 12 balloons and 3 children. How many balloons should we give each child, so they each have a fair share?	Count 12 objects. Count 3 to represent the 3 children. Deal out the 12 objects one at a time until they are all out. Count how many objects are in each place.

TRY IT

Practice determining if students are using Counting Strategies by giving students the problems to solve in Tables 3.1 and 3.2.

Have objects available for students to use, but do not suggest using them. Ask to hear student thinking. Watch to see how they use objects if they do.

It's fantastic for very young students to make sense of these problems and use Counting Strategies to solve them, but it quickly becomes laborious and inefficient as the numbers grow. More importantly, if students are only rewarded for finding answers counting one by one and never encouraged beyond it, then they might not develop the more complex additive and multiplicative ways of solving such problems.

Such stranding in the world of Counting Strategies is problematic enough within the bounds of basic arithmetic. The reality is that if left to use only this least sophisticated of reasonings, students will have no recourse but to rely solely on Counting Strategies into algebra and beyond. No mnemonic trick or algorithm can address this fundamental limiter on speed, accuracy, and—most crucially for learning future content—understanding.

Consider this example of higher math where students can get correct answers using Counting Strategies—and hate every minute (of which there will be many) doing it.

What is involved for a student relegated to counting when asked to find the distance between two points (2, –8) and (–3, 4) in late middle school/early high school?

We give them a formula like this:

$$(x_2 - x_1)^2 + (y_2 - y_1)^2 = c^2$$

Notice how they can figure each bit by counting. If you are only looking at the answer they get, you'd never know.

$$((-3) - 2)^2 + (4 - (-8))^2 = c^2$$

TIP

How can you tell if your students are counting one by one to solve a problem? If they put up their fingers one by one, that's a definite sign. Younger students may move counters or draw tally marks. There may be more subtle clues, like nodding their heads or moving their lips several times repetitively. When in doubt, remember that nothing can substitute for asking a student to describe their thinking.

The student finds $(-3) - 2$ by Counting Back by ones to -5: -3. $-4 \rightarrow -5$

$$(-5)^2 + (4 - (-8))^2 = c^2$$

Then the student finds $4 - (-8)$ by mimicking a rule "minus, minus, plus, plus" to get $4 + 8$ and counts by ones to get 12. $4 \rightarrow 5 \rightarrow 6 \dots 11 \rightarrow 12$

$$(-5)^2 + (12)^2 = c^2$$

Then they have to do $(-5)^2$. Notice the pit of dread that just formed in your stomach from the very idea of solving that via counting. Empathize.

$$(-5)^2 + (12)^2 = c^2$$

For teachers of older grades, here is the tedium/irrelevance/boredom in which your counting students live; for lower grades teachers, this is why algorithms are insufficient preparation for higher math.

FREQUENTLY ASKED QUESTIONS

Q: There are more detailed, parsed-out levels of Counting Strategies (Carpenter et al., 2014). Why don't you go into them here? Should we have students move through those levels?

A: I don't advocate for spending a lot of time and effort helping students move to a different kind of Counting Strategy. The big goal here is to encourage and support students in figuring out what problems are asking and help students solve them using Counting Strategies. As soon as students are solving problems using Counting Strategies, the goal is to help them begin to think in bigger jumps of numbers than one at a time—the goal is to begin to develop Additive Reasoning.

WHAT IS ADDITIVE REASONING?

Unlike the previous Counting Strategies, where students count one by one, Additive Reasoning is about solving the same problem using bigger jumps of numbers.

Join Holly and her fourth-grade students in a subtraction Problem String.

Holly starts. "We've been working on subtraction recently," she says, "and we've been talking about two different ways that we can subtract. We could either remove or find the difference. Can somebody say something about that? Victoria?"

Victoria answers, "Finding the difference is that you get the smaller number and then you add some numbers together to get other numbers."

Holly asks, "Okay, when would we find the difference? Sophia?"

Sophia replies, "If the numbers are pretty close."

"If they're pretty close, it's easy to find the difference," answers Holly as she moves her hands close together.

Holly asks, "When would it make sense to remove?"

Students call out, "When they're far."

Holly replies, "When they're far apart," as she moves her hands far apart.

Holly gives the first problem, 68 – 39, and circulates to see what relationships students are using.

Back at the board, Holly asks, "All right, Christian, what'd you get?"

When Christian responds, "29," Holly follows with, "And how did you get 29?"

Christian's answer: "I subtracted."

Rather than assume, and to help all students follow Christian's thinking, Holly asks, "Where'd you start?"

Christian answers, "With 68. And I subtracted 30."

Holly, knowing that it's important to highlight where that lands because otherwise students might just start jumping haphazardly, without considering why they are making a jump, asks, "And that got you to—?"

As Christian keeps explaining, Holly represents his thinking on the board: "38. And then I subtracted 8. That got me to 30. Then I had one more, so I subtracted 1. I got to 29."

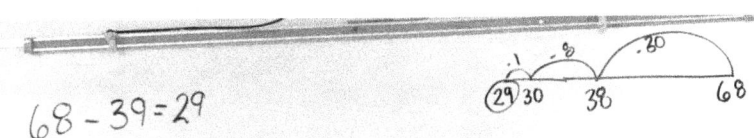

$68 - 39 = 29$

Holly continues by calling on Sophia. "Okay, Sophia. You found the difference, I think. How did you?"

Sophia starts. "I added 1 so I will get to 40."

Again, Holly helps the rest of the class follow along. "So where'd you start?"

When Sophia answers, "39," Holly draws a number line starting at 39.

$68 - 39 = 29$

39

Holly says, "So you started at 39. Okay. And you—?"

Sophia responds, "Added 1 so I could get to 40. And I added 8, so I wouldn't have to put 8 at the end."

Holly again asks for a landing spot. "Okay, so you added 8 and that got you to—?"

Sophia says, "48. And I just added 20. That's 68. And all of that equals 29."

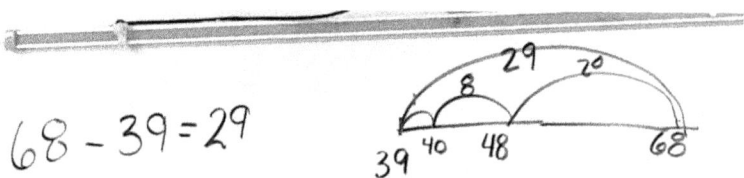

$$68 - 39 = 29$$

Holly gives the next problem, 67 – 38, and suggests, "Let's find the difference. We could remove, we could find the difference. Let's find the difference."

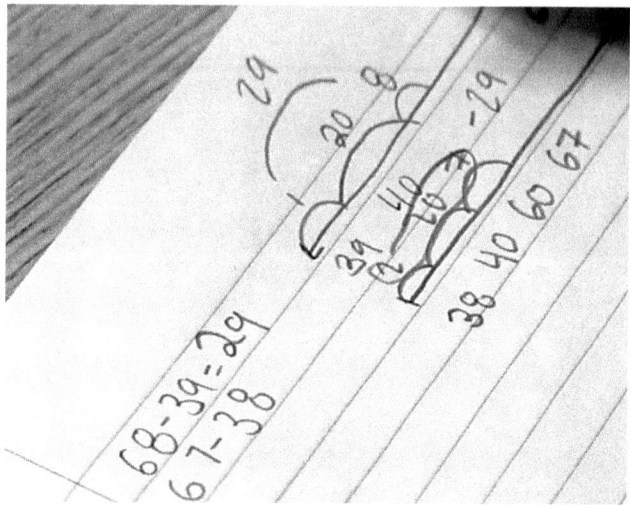

After she circulates, Holly calls on Gilberto, who says, "I made the difference. I started off with 38."

Holly holds her hands up to the 39 and 68 from the previous number line and says, "So here's where I have 39 and 68. I'm just going to," and she motions to the left a bit, "okay—38," as she draws the 38 just to the left of the 39. She motions to draw students' attention to the fact that she is drawing the number lines in relation to each other. The 38 should be to the left of the 39 on the number line above.

Holly listens intently as Gilberto explains his strategy and Holly represents it in one big jump. This is purposeful because her goal is to get students to notice the relationships between the problems.

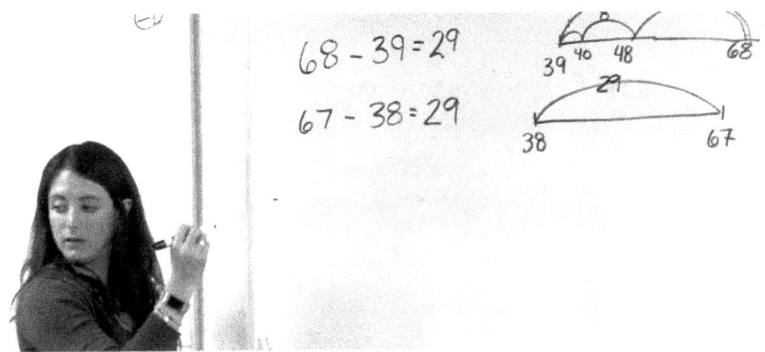

The String progresses through these problems: 63 – 34, 61 – 32, 69 – 40. Each time Holly represents student thinking as distance, Holly motions with her hands at the board as students helped her decide where the numbers for each number line would appropriately be written.

After the 61 – 32 problem, Holly asks Sophia why she is smiling.

"Because all of our answers have been 29," answers Sophia.

After representing 69 – 40, Holly puts her hands up to the numbers to represent the distance of 29, shifts her hands around, and asks, "If you had to choose one of these problems to do, which one would you choose?"

When students say the 69 – 40 problem, Holly pushes, "Why? You kind of explained there's a friendly number in it. And you were saying something about how they're all 29. Alan, do you have any thoughts about that?"

Alan talks about how the numbers in the subtraction problems are all close to each other. Holly brings more voices in. "Things are kind of moving around, aren't they? But what's happening to the distance between them?"

When several students call out, "Stays the same!" Holly says, "It's staying the same. Did you know you could do that? You could just shift problems down a little bit or up a little bit and the distance would stay the same? It's interesting, isn't it? So, I wonder, what if I gave you a problem like this? 126 − 97. Could you shift that problem at all? To make it a little bit easier?"

Watch Holly's class reason additively

https://qrs.ly/
5xg3rea

To watch this clip and see these students reasoning in action, see the QR code on this page.

These students are reasoning additively, building notions of place value, magnitude, equivalence, and the two meanings of subtraction. They are using what they know to solve problems and in the process developing more sophisticated strategies.

Additive Reasoning can be characterized by two different aspects. One is the ability to solve problems by moving (thinking) in jumps larger than one at a time. The other is to conceive of multidigit numbers as a single decomposable whole as opposed to a list of digits.

Moving in jumps larger than one at a time requires:

- Considering a multidigit number as a single whole representative of one value, as opposed to a list of individual digits. For example, 53 is a unit, one 53 and simultaneously 53 ones, but also as 50 and 3, not the digit 5 and the digit 3.

This is distinct from a multiplicative understanding of place value, where 53 is $5 \times 10 + 3 \times 1$.

- The simultaneity to consider more than one number in the above manner at a time, so that a problem like 53 + 28 isn't 53 ones and then 28 ones, but one 53 and one 28 and simultaneously 50 and 3 and 20 and 8 but also 53 and 20 and 8 or 53 and 30 minus 2.

Use of algorithms such as the North American Traditional Addition Algorithm directly undermines both of these key abilities. By isolating all operations to single-digit arithmetic, practice using the algorithm is never practice thinking additively about numbers bigger than single-digits.

Admittedly, a small percentage of students will develop Additive Reasoning *despite* algorithm-focused teaching. As teachers, we all want to do a lot better than having students learning *despite* our efforts.

TIP

How can you tell if a student is using Additive Reasoning? If the problem calls for addition or subtraction and the student does not count one by one, but thinks in terms of bigger jumps than one by one, chances are high the student is using Additive Reasoning.

What's the Difference Between Counting Strategies and Additive Reasoning?

48 + 5

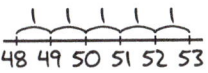

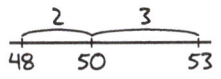

48 + 2 = 50
50 + 3 = 53

The student is counting one by one to solve the problem, 48 + 5, shown by individual jumps on an open number line.

The student uses bigger jumps to solve the same problem, 48 + 5, shown on an open number line and with equations.

A student demonstrates Additive Reasoning when adding 995 + 457, by taking 5 from the 457 to give to the 995 to make 1000 + 452.

Video of a student solving an addition problem using a Counting Strategy

https://qrs.ly/qvg3reb

Video of a student solving an addition problem using Additive Reasoning

https://qrs.ly/dog3ree

Source: istock.com/Goodboy Picture Company

Additive Reasoning is also shown more subtly in Table 3.3 for problems like $2.9 + 4.6$, $47 + 38$, $4\frac{1}{2} + 9\frac{3}{4}$, and $995 + 457$.

TIP

Notice that the answer to $995 + 457$ when giving 5 from 457 to 995 to get $1000 + 452$ actually reads aloud the same as the problem if you substitute the word *and* for *plus*. Read aloud 1000 and 452 and it's the same as if you say 1452. It's almost like the question is the answer. What is 1000 and 452? Yep—1452. When the question is the answer, this is often a sign that you were involved in reasoning, not mimicking.

TABLE 3.3 ● Major Strategies for Addition

$2.9 + 4.6 = 7.5$	4 0.6 — 2.9 6.9 7.5	An Add a Friendly Number strategy
$47 + 38 = 85$	3 35 — 47 50 85	A Get to a Friendly Number strategy
$4\frac{1}{2} + 9\frac{3}{4} = 14\frac{1}{4}$	10 $-\frac{1}{4}$ — $4\frac{1}{2}$ $14\frac{1}{4}$ $14\frac{1}{2}$	An Add a Friendly Number Over strategy
$995 + 5 = 1000$ $+ 457 - 5 = + 452$ 1452		A Give and Take strategy

Video of a student solving subtraction problem using a Counting Strategy

https://qrs.ly/ p8g3reg

The four strategies shown here represent the major relationships and strategies that students need to develop for Additive Reasoning with addition. There is a precursor fifth strategy that we promote with young students called Splitting by Place Value.

With subtraction, a student demonstrates Additive Reasoning when finding $72 - 38$ by removing too much, 40 to get $72 - 40 = 32$, and then adding back the extra 2, for 34. Additive Reasoning is also shown in Table 3.4 for the problems $4.6 - 2.8$, $72 - 38$, $7\frac{1}{4} - 4\frac{1}{2}$, and $5621 - 1989.5$.

Video of a student solving a subtraction problem using Additive Reasoning

https://qrs.ly/ 48g3rei

TABLE 3.4 ● Major Strategies for Subtraction

$4.6 - 2.8 = 1.8$	-2.2 -0.6 — 1.8 4 4.6	A Remove to a Friendly Number strategy
$72 - 38 = 34$	$+2$ -40 — 32 34 72	A Remove a Friendly Number Over strategy

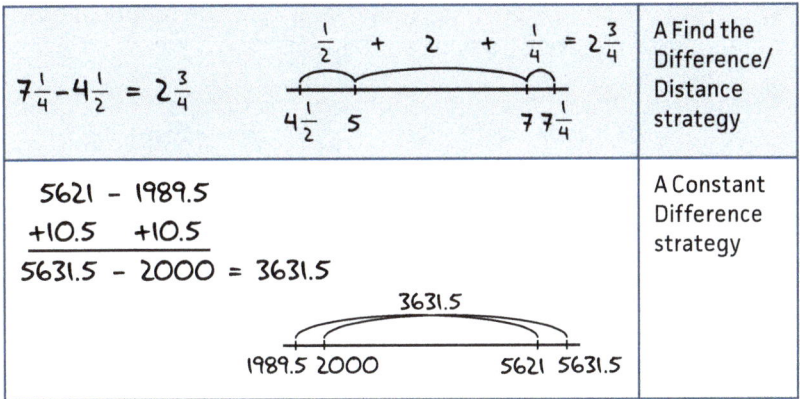

$7\frac{1}{4} - 4\frac{1}{2} = 2\frac{3}{4}$ $\frac{1}{2} + 2 + \frac{1}{4} = 2\frac{3}{4}$ $4\frac{1}{2}$ 5 7 $7\frac{1}{4}$	A Find the Difference/ Distance strategy
$5621 - 1989.5$ $+10.5 \quad +10.5$ $5631.5 - 2000 = 3631.5$ 3631.5 $1989.5 \ 2000 \qquad 5621 \ 5631.5$	A Constant Difference strategy

The four strategies shown here represent the major relationships
and strategies that students need to develop for
Additive Reasoning with subtraction.

FREQUENTLY ASKED QUESTIONS

Q: Where can I learn more about Additive Reasoning?

A: To learn more about Additive Reasoning and to see it in action, check out the module on Additive Reasoning in my online workshop, *Developing Mathematical Reasoning*, at https://www .mathisfigureoutable.com/dmr/workshop, as well as my upcoming books *Developing Mathematical Reasoning K–2* (available 2025) and *Developing Mathematical Reasoning 3–5* (available 2026).

Developing
Mathematical
Reasoning Online
Workshop

https://qrs.ly/
6rg3rek

The examples in the previous section each show reasoning in bigger jumps of numbers than counting one by one. The traditional algorithm for addition can trap students into looking like they are multidigit additive reasoners, when they are actually using Counting Strategies, or better but not good enough single-digit additive strategies.

If the goal of mathematics education involves learning math and learning *to math*, endlessly practicing counting does not support that goal.

THE TRAP OF A TRADITIONAL ADDITION ALGORITHM

In the United States, textbooks have typically chosen a particular multidigit addition algorithm. The steps of that procedure can be summarized as follows: Line up the numbers by place value. Add the columns of digits starting with the smallest place value. Regroup (carry) as needed. Repeat.

> *In this book, we will analyze the traps of the traditional algorithm used in the United States. Other digit-oriented addition algorithms have the same or similar traps.*

While seemingly straightforward, this series of steps has several traps.

Let's walk through a problem like 26 + 49, considering how students could get correct answers while trapped by the algorithm into not reasoning at all about 26 or 49 or the sum of 26 and 49.

To begin, the student lines up the numbers, treats the problem as columns of digits, and starts with the right column (the smallest components of the addends).

STEP 1: ADD THE 6 + 9

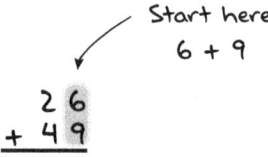

STEP 1, TRAP 1: THE DIGIT TRAP AND IDENTITY TRAP

Starting with the 6 and 9 from the 26 and 49 works against student intuition. Students who have not been introduced to algorithms almost always approach solving multidigit problems left to right (biggest part of the numbers first) instead of right to left (smallest part of the numbers first). This intuition has critical implications for the understanding of place value and ought not be undermined. Students enter multidigit problem solving with the correct intuition that the leftmost digits make the biggest difference in the magnitude of the final answer. Not all student intuition is correct, but when it is correct, undermining it is tragic. Nothing destroys confidence or trust faster.

Picture yourself at an age when you are learning to add numbers like 26 and 49. Students at that stage are thinking about the 20 and 40 far more than the 6 and 9. These students should be learning to think about the 26 as: 20 and 6, one more than 25, one less than 27, four less than 30, more than the teens (11–19) and less than 100. All that meaning is destroyed, or at least severely limited, when a number is viewed as a list of digits (which the algorithm requires) rather than as one whole number.

The algorithm demands that students not only think about 26 as the digits 2 and 6; it requires they start with the most inconsequential part of 26, the 6, and then add it to the most inconsequential part of 49, the 9. Students who are intuitively thinking about the big numbers have the potential in this moment to think, "Wait, what? Start with the 6 and 9? I'm thinking about the 20 and 40. Weird. I guess I'm odd. Everyone around me is dutifully doing what the teacher says, so I guess it makes sense to them. It doesn't make sense to me. I guess they are math people. I guess I'm not a math person."

The nonintuitive approach of algorithms can trap a student into thinking they are not a math person.

STEP 1, TRAP 2: STUCK IN COUNTING STRATEGIES

Following the steps of this algorithm allows students to use Counting Strategies and never build their brain's capacity to reasoning additively.

TIP

Intuition dictates that you consider the bigger part, the left-hand side, of the number first. To check this for yourself, think of your online bank account. You're going to pay a bill that is $247.53. You need to make sure there is enough in your account to cover that payment. What are you thinking about right now? The $0.53? Or the roughly $250? Yep. The bigger numbers matter more and demand your attention.

Not every student will get stuck in Counting Strategies, and those who do will not get stuck with every algorithm. But when students do happen to learn to reason while being taught an algorithm, this reasoning learning is despite what the algorithm requires, not because of it.

After lining up the numbers, the student does one of the following to find the sum of 6 and 9:

- Student A: Count 6, count 9, count all of them to 15. *Counting 3 Times strategy.*

- Student B: Start with the first number, 6, and Count On by ones: 6. 7 → 8 → 9 → 10 → 11 → 12 → 13 → 14 → 15. *Counting On from the first number strategy.*

- Student C: Start with the higher number, 9, and Count On by ones: 9. 10 → 11 → 12 → 13 → 14 → 15. *Counting On from the higher number strategy.*

It "works"! Students get the correct answer to the first column addition. They are not using Additive Reasoning. They are not building a sense of addition. They are not practicing addition. They are practicing counting. Even though the third student used the most sophisticated Counting Strategy, this is not sophisticated enough. It is not Additive Reasoning.

Consider that if counting is sufficient to the ends of a mathematics education, there is no reason to not always use a calculator after first grade. Students have got counting by that point, or if they do not, there is a multitude of better ways to teach it than stacking progressively more opaque algorithms on top of each other.

STEP 1, TRAP 3: STUCK IN SINGLE-DIGIT ADDITIVE REASONING

You might be thinking, "I do not need to worry about my students being trapped in Counting Strategies. My students are not counting one by one; they are reasoning additively about those columns."

This is true for many (but definitely not all) students. However, even if students have progressed to additive strategies with single-digits despite algorithm-focused instruction, the algorithm still limits their exposure to more sophisticated thinking.

A student trapped in the algorithm could think additively about those single digits. To find the sum of 6 and 9:

- Think 6 to 10 is 4, so then 5 more for 15, using a Get to 10 strategy.

6 + 9 using a Get to 10 strategy

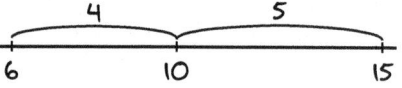

- Think 6 + 6 is 12, so 6 + 9 is 3 more, 12 + 3 = 15, using a Doubles strategy.

6 + 9 using a Doubles strategy

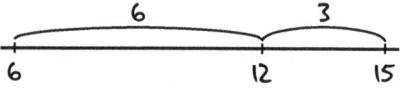

- Think 6 + 10 is 1 too much, so 16 − 1 = 15, using an Over strategy.

6 + 9 using an Over strategy

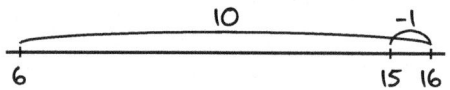

These three strategies are better than Counting Strategies; they show more sophisticated thinking. But it's still only single-digit Additive Reasoning. Students stuck here in the algorithm will never have any practice thinking additively about values larger than 18 (the largest sum of two single-digit numbers). To use metaphors again, this could be similar to being limited to 2-lb weights no matter the exercise. It doesn't matter how many reps you do, 2 lbs will not adequately prepare you for 20 lbs. Nor will only walking around in a wading pool ever be sufficient preparation for going scuba diving.

Single-digit Additive Reasoning is not enough preparation for Multiplicative Reasoning. It is necessary, but not sufficient.

STEP 2: WHAT TO DO WITH THE 15

Now, the student has to handle the 15 they just found in the first column. The algorithm requires the student now to break up the 15, leaving part in the ones column and putting the other in the tens column.

So 6 + 9 is 15,

Carry the 1 $\longrightarrow$ 1

2 6

Put 5 here + 4 9

$\longrightarrow$ 5

This can be so confusing to students that teachers have come up with nonsense like this:

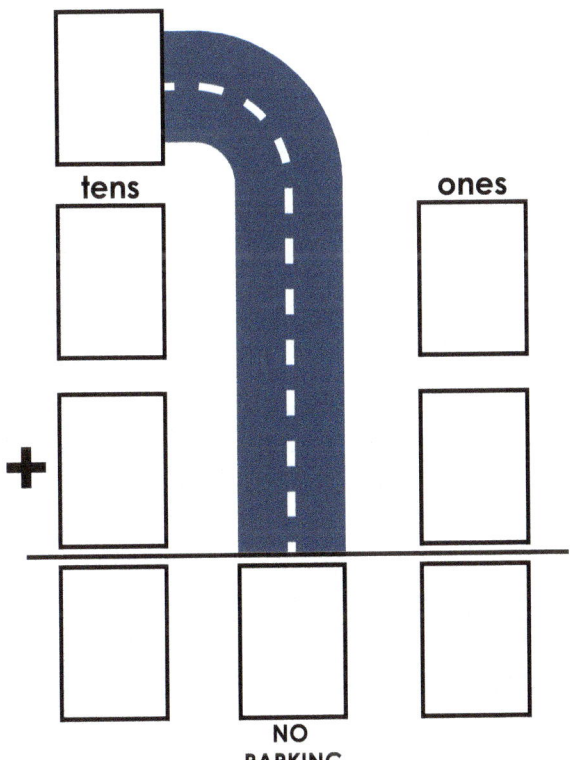

tens   ones

+

NO
PARKING

I'm using the word *nonsense* very particularly here. The sense the graphic is trying to make, the idea that a 1 is carried over by a car, has absolutely nothing to do with mathematical reasoning. It's not even tangential. It's completely nonsensical.

Remember, if we are using a song or rhyme to remember something, it is at best vocabulary, and at worst replacing a mathematical relationship whose absence will make everything else harder. If we are trying to make it cute, we are probably in fake math mimicking.

STEP 2, TRAP: STUCK IN "CARRYING/REGROUPING" USING A DIGIT FOCUS

The student takes the 15 from the sum of the first column, and thinks:

- The 15 looks like 1 and 5, put the 5 down there and the 1 up there, never thinking about more than digits. No place value. This is terrible. Numbers are more than mere collections of digits. If a student stays here, they may very well accidentally write down the 1 in the ones column and the 5 in the tens column because when you write 15, which number do you typically write first?

So 6 + 9 is 15. When I write 15,
I write the 1 first, then the 5.

Carry the 5 ⟶ 5
Put 1 here   2 6
                       + 4 9
                   ⟶ 1

TIP

Consider that for many students, the number 15 is tricky to write because you say the "five" part of the number first: "five-teen." But when you write, you write the 10 part first, then the 5: 15. The algorithm then reverses that—write the 5 first and then the 1. No wonder students struggle!

- 15 is made of 10 and 5; put the 1 that stands for 10 up there and the 5 down there because that's what you do, not realizing the place value at work. The student is using place value but focusing on digits. This is slightly better. Barely.

- 15 is made of 10 and 5; put the 1 that stands for 10 with the other tens and the 5 with the ones, using place value understanding. This is the best-case scenario for students inside this algorithm. Some students get this. Most do not.

- While this in itself is not a trap, the fact that a student is forced to break up 15 as 10 and 5 inside of 26 + 49 can trap students into looking like they are reasoning additively about 26 and 49 while in reality they are reasoning about smaller numbers. This can mean these students are never invited into the opportunity to grapple with larger numbers. They never increase the mass of the weights they are lifting. They never swim farther than across the wading pool or do more than press down one guitar string at a time.

TIP

A general approach that generally doesn't work is not a good general approach.

STEP 3: WHEN THE STUDENT MOVES ONTO THE NEXT COLUMN, SOME OF THE SAME THINGS HAPPEN AGAIN

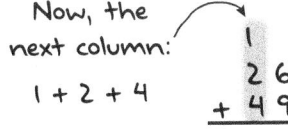

Now, the next column:

1 + 2 + 4

STEP 3, TRAP 1: STUCK IN COUNTING STRATEGIES, AGAIN

- Count 3 Times
- Count On from either number by ones

Again, this student is not building or using Additive Reasoning, just getting answers.

Source: istock.com/PeopleImages

STEP 3, TRAP 2: STUCK IN SINGLE-DIGIT ADDITIVE REASONING, SECOND VERSE, SAME AS THE FIRST, BUT SO MUCH WORSE BECAUSE IT'S DESTROYING PLACE VALUE RATHER THAN MERELY IGNORING IT

- 4 + 2 is 6 and 1 more is 7.

Again, this is better than a Counting Strategy, but the student is not thinking about the magnitudes involved, 40 and 20 and 10. That can be advantageous once a person has built Additive Reasoning, but it's not great to be trapped here.

Now that you've found the sum of the left column to be 7, what might you do with it?

$$
\begin{array}{r}
1 \\
2\ 6 \\
+\ 4\ 9 \\
\hline
7\ 5
\end{array}
$$

1 + 2 + 4 is 7, write down the 7

TRAP 3: BEING STUCK IN A DIGIT-ORIENTED APPROACH TO THE ANSWER

- I got 7, put it next to the 5, so the answer is "7" "5", not acknowledging that 7 in the tens is 70, the answer is just those digits 7 and 5 next to each other. This digit approach traps students into never really acknowledging the number 75.

- I got 7, put it next to the 5, and 7 next to 5 means 75, doing the steps with digits, but then coming back to place value. This can almost be a "do the thing, then think about it" approach. Better, but not having the opportunity to grapple with 70 and 5 being 75.

- 10 + 20 + 40 is 70, but there's that 5 already there, so 70 + 5 is 75, thinking about and using place value. This is the best-case scenario, but if this is all the student has been encouraged/taught/given the opportunity to do, they'll just keep adding larger numbers and having the same results.

- Notice that for each of these three finishes, the student *reads off the answer*, 75. This reading off behavior is different than being cognitively involved in the relationships during the solution process. A student who reads off the answer as the last step could almost be surprised at the 75 because they weren't thinking about the magnitudes; they were thinking about digits. A student who is thinking about 26 and 49 as 26 and 50 then back up 1 is cognitively involved in the sizes of the numbers the whole time (Klein et al., 1998).

It "works"! The student using the algorithm got the correct
answer to the problem. If they are using a Counting Strategy,
they are practicing counting. Even in the best-case scenario,
they are at most using single-digit Additive Reasoning. This
is not sophisticated enough. It is not sufficient for developing Additive Reasoning with subtraction or moving on to
Multiplicative Reasoning. It is the equivalent of building a
house on tissue paper in terms of preparing for Proportional
and Functional Reasoning.

Many of the traditional algorithms share these three general traps:

1. The less sophisticated reasoning trap—the trap of being
   able to use reasoning in a prior level(s) and get correct
   answers, therefore never learning the reasoning that kind
   of problem could build and needs to build in order to then
   build on it. For the addition algorithm, this is the trap of
   using Counting Strategies instead of building Additive
   Reasoning.

2. The digit trap—the trap of treating numbers like lists of
   digits and therefore not grappling with the place values/
   magnitudes involved. So many answers are unreasonable
   because the student was in the mode of repeating steps
   with digits. Students read off the answer because they were
   not cognitively involved in the relationships.

3. The false definition of math trap—the trap of believing
   the myth that math-ing means mimicking. Specifically,
   the trap that addition means: Line the numbers up.
   Start with the smallest column. Split up the answer. Put
   part in the other column. Repeat. "Addition" becomes a
   thing to do, not an operation to understand/feel/intuit
   with. If I am not good at rote-memorizing and mimicking,
   or I refuse to do so because I think I should be able to
   understand, my identity as a math-er takes a hit. *I don't
   understand/can't memorize these steps—guess I don't have the
   math gene.*

TRY IT

What could it look like to reason additively about 26 + 49?

- Think about 26 + 4. How could you use 26 + 4 to think about 26 + 49?

If 26 + 4 is 30, what's left?

$$\begin{array}{r} 26 \ + 4 = 30 \\ + 49 \ - 4 = +45 \\ \hline 75 \end{array}$$

- Think about 26 + 40. How could you use 26 + 40 to think about 26 + 49?

If 26 + 40 is 66, what's left?

- Think about 26 + 50. How could you use 26 + 50 to think about 26 + 49?

If 26 + 50 is 76, what's left?

FREQUENTLY ASKED QUESTIONS

Q: What about models like tens frames and hundreds charts?

A: Both ten frames and hundreds charts are fine models to *build mathematical relationships*. The 5- and 10-structure of a ten frame is great to help students to begin to see and use numbers as 5—*and* in relation to 10. For example, 4 is 1 less than 5, 7 is 2 more than 5, 8 is 2 less than 10, and 13 is 10 + 3. Likewise, the structure of hundreds charts is great to help students make sense of multiples of 10 and how 44 relates to 34 and 54 but also 43 and 45.

However, neither of these models is good as a *tool for computing* because students get stuck in counting by ones (Counting Strategies) when they should be building Additive Reasoning. Better tools for computing are Number Racks and the open number line because they allow students to reason with bigger jumps than one at a time.

HOW ARE YOU THINKING ABOUT SUBTRACTION RIGHT NOW?

Was your subtraction instruction primarily algorithms? How do you think about subtraction problems now?

In this section, focus on the way your brain handles these problems. Solve the problem, then read the descriptions underneath, and choose one that best fits your thinking.

How do *you* think about: 42 – 6, 367 – 98?

42 – 6

If you thought 42. 41 → 40 → 39 → 38 → 37 → 36 (not picturing or writing numbers with arrows, but counting one by one, whether in your head or somewhere else), then you were using a Counting Strategy.

If you thought 42 down 2 is 40, down 4 more is 36, then you were using an Additive Strategy.

367 – 98

If you lined up the digits in columns and thought 7 – 8. Can't do that, so regroup (borrow) to make it 17 – 8. 17. 16 → 15 → 14 → 13 → 12 → 11 → 10 → 9. And then you continued to do that in the next column, you were using a Counting Strategy.

If you thought 17 – 8 is like 17 minus 7 is 10 so down one more is 9, and when you got to the next column you thought about 15 – 9 as 15 minus 10 is 5 but you removed too much so 15 – 9 is 6, then you were using teens minus single-digit Additive Reasoning.

If you thought 367 – 98 is really close to 367 – 100 which is 267, but you removed too much, so you adjusted up by 2 so 269, then you were using Additive Reasoning about subtraction with multidigit numbers.

THE TRAP OF A TRADITIONAL SUBTRACTION ALGORITHM

"Pam, we know you work with secondary teachers, but could you do a math-teaching workshop for our elementary teachers?"

"Yes!" I respond. This request comes early in my career, when I have started diving into research and my kids' elementary classrooms. I have so many helpful things to share.

"Great. We'd really love it if you could address subtraction across zeros," the teacher leader requests.

Not a thing I have ever heard of—though I can guess—so I ask her to clarify. "What does that mean?"

"You know, problems like 100 – 98 or 1000 – 997. They are so hard for kids!" she replies.

This is one of my first experiences when I clearly see teachers and students trapped in the faulty notion that students need to be able to repeat complicated series of steps to solve problems, even problems that are actually so figure-out-able.

TRY IT

Use the Find the Distance strategy to solve 100 – 98 and 1000 – 997. Ask students to think about how far apart the two numbers in each problem are. Listen to their reasoning. These numbers are so close together, of course finding the distance is nice for these numbers. How far apart do you think the two numbers should be before this Finding the Distance/Difference strategy is not such a good strategy?

In the United States, textbooks have typically chosen a particular multidigit subtraction algorithm. The steps of that procedure can be summarized as: Line up the numbers by place value. Subtract the columns of digits starting with the smallest place value. Regroup (borrow) as needed. Repeat.

In this book, we will analyze the traps of the traditional algorithm used in the United States. Other digit-oriented subtraction algorithms have the same or similar traps.

While seemingly straightforward, this series of steps has several traps.

Let's walk through a problem like 83 – 27, considering how students could get correct answers while trapped by the algorithm into not reasoning at all about 83 or 27 or the removing of 27 from 83 or the distance from 27 to 83.

To begin, the student lines up the numbers, treats the problem as columns of digits, and starts with the right column (the smallest components of the addends).

STEP 1: FOR 83 – 27, LINE UP THE DIGITS AND WORK WITH THE SMALLEST NUMBERS FIRST

For 83–27, line up the digits.

```
  8 3
- 2 7
```

Start here
3–7

```
  8 3
- 2 7
```

STEP 1, TRAP 1: DIGIT TRAP AND IDENTITY TRAP

Just as with the addition algorithm, students are required to treat the numbers as lists of digits: in this case, 8 and 3 and 2 and 7, not as entities 83 and 27. They are not thinking about *80 something* subtract *20 something* or even *80 and a bit* subtract *almost 30*. No, they are lining up the digits and then thinking about 3 – 7, the smallest components of this problem. If we have not forced these algorithms, students' intuition would be to deal with the big numbers first, thinking about 80 and 20, not 3 and 7. Forcing students to consider 3 – 7 first works against that intuition. This sends the message: *Don't think in math class; just mimic. If that doesn't make sense to you, you're not a math person.*

STEP 1, TRAP 2: YOU CAN'T SUBTRACT A NUMBER FROM A SMALLER NUMBER

Students are told to first subtract 7 from 3. But what happens when 7 is larger than 3? Young students are told, when you *cannot* subtract, you must regroup (borrow) from the next place. A quick online search shows many cute versions of this ditty:

- "More on the floor? Go next door and borrow some more."
- "More on the top? There's no need to stop."
- "Numbers the same? Zeros the game."

You might have heard this alliterative memory aid: "bigger bottom, better borrow."

For 83–27, line up the digits.

```
  8 3
- 2 7
```

Start here
3–7

```
  8 3
- 2 7
```

More on the floor?
Go next door,
borrow some more.

or

Bigger bottom,
better borrow.

Teachers of older grades cringe because when the second number is larger than the first in a subtraction problem (more on the floor), you can still subtract! For this problem, 3 – 7 is –4. When students reach middle school and operations with integers (positive and negative numbers), they must either (1) ignore what they learned in the subtraction algorithm, or (2) compartmentalize—when subtracting whole numbers and decimals, remember you cannot subtract a number from something smaller. When doing middle school math with integers, you *can* subtract any number from another number. To students attempting this compartmentalization, math is inconsistent and certainly not figure-out-able.

This is a perfect example of a rule that expires. Karp, Bush, and Dougherty describe these as "rules that seem to hold true at the moment, given the content the student is learning. However, students later find that these rules are not always true; in fact, these rules 'expire.' Such experiences can be frustrating and, in students' minds, can further the notion that mathematics is a mysterious series of tricks and tips to memorize rather than big concepts that relate to one another" (Karp et al., 2014).

STEP 1, TRAP: THE FALSE DEFINITION OF MATH/IDENTITY TRAP

As students are told to do these steps, the math/identity trap is looming large. Students learn to follow steps and think that subtraction means: *Sing the ditty, do steps that don't make sense, get an answer, and be done.* When students don't rote-memorize well or refuse to blindly follow steps, students often think, "I guess I'm bad at subtraction. I don't have the math gene."

STEP 2: 83 – 27. SINCE YOU "CANNOT" SUBTRACT 3 – 7, REGROUP (BORROW) FROM THE 8

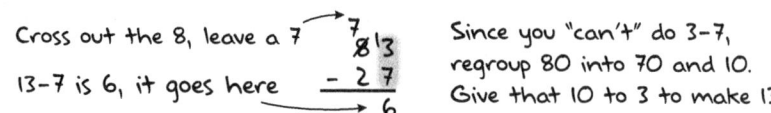

Cross out the 8, leave a 7

13-7 is 6, it goes here

Since you "can't" do 3-7, regroup 80 into 70 and 10. Give that 10 to 3 to make 13.

STEP 2: TRAP 1 STUCK IN COUNTING STRATEGIES

After the regrouping (borrowing), the student does one of the following to find 13 – 7:

- Student A: Count 13, remove 7, count the leftover 6. *Counting 3 Times strategy.*

- Student B: Start with 13 and Count Back by ones: 13 → 12 → 11 → 10 → 9 → 8 → 7, so the answer is 6. *Counting Back strategy.*

- Student C: Start Counting Back with 12 → 11 → 10 → 9 → 8 → 7 → 6, so the answer is 6. *Counting Back strategy.*

It "works"! Students get the correct answer to the first column subtraction. They are not using Additive Reasoning. They are not building a sense of subtraction. They are not practicing subtraction. They are practicing counting. Even though Students B and C used the most sophisticated Counting Strategies, this is not sophisticated enough. It is not Additive Reasoning.

Students are trapped in Counting Strategies, getting correct answers to subtraction problems but not growing mathematically.

STEP 2, TRAP 2: STUCK IN TEENS MINUS SINGLE-DIGIT ADDITIVE REASONING

You might be thinking, "I do not need to worry about my students being trapped in Counting Strategies. My students are not counting one by one; they are reasoning in bigger jumps as they subtract in those columns."

This may be true for many (but definitely not all) students. However, even if students have progressed to Additive Strategies in the column subtractions despite algorithm-focused instruction, the algorithm still limits their exposure to more sophisticated thinking.

TIP

Researchers have helped us understand that there are different Counting Back strategies students could use. Finding which Counting Back strategy students are using is not important, because the emphasis is on recognizing that students are Counting Back by ones and helping them develop more sophisticated Additive Strategies.

Source: istock.com/vejaa

A student trapped in the algorithm could think additively about 13 – 7. To find 13 – 7:

- Think 13 subtract 3 is 10, so subtract 4 more, so 6, using a Remove to 10 strategy.
- Think 7 + 7 is 14, so 14 – 7 is 7, so 13 – 7 is 6, using a Doubles strategy.
- Think 13 – 10 is 3, but that is 3 too much, so adjust 3 up to 6, using an Over strategy.

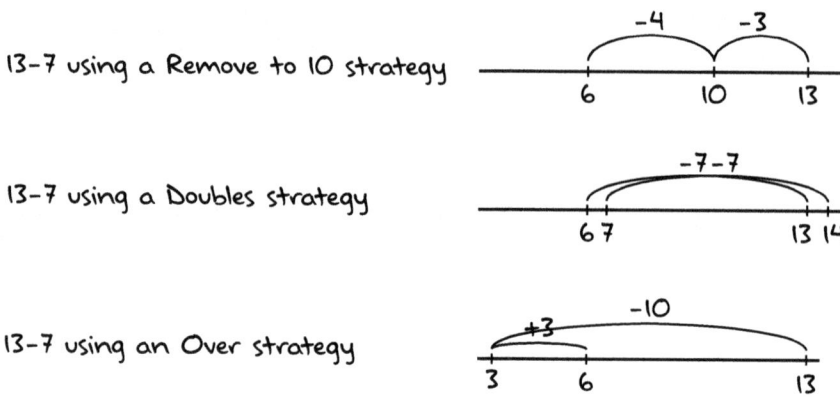

13-7 using a Remove to 10 strategy

13-7 using a Doubles strategy

13-7 using an Over strategy

These three strategies are better than Counting Strategies; they show more sophisticated thinking. But it's still only small number subtraction Additive Reasoning.

Always solving subtraction problems that require regrouping, where "More on the floor, go next store, borrow some more" with the algorithm means students will end up with at most a teen number subtracting a one-digit number. That means that stuck in this algorithm, students will never subtract anything larger than 18 – 9. "That's fantastic!" algorithm proponents cheer.

"Students only need to rote-memorize addition and subtraction within 20 and—voilà! They can solve any subtraction problem, no matter how large the numbers are."

Yet then we are left with students who *never* grapple with subtraction problems with numbers larger than 18 – 9.

Students need experience grappling with larger numbers in subtraction problems. This grappling builds a sense of friendly numbers, place value, reasonableness, and magnitude. Why are students' answers to subtraction problems often so unreasonable? We took them out of reasoning and into mimicking, therefore depriving them of opportunities to make sense of the numbers.

TIP

It's 18 – 9 because if the digit is 9, you can subtract any other digit and don't need to borrow. If that digit is 8, you regroup to make it 18 and subtract 9. Every other combination of digit subtract digit consists of smaller digits.

STEP 2, TRAP 3: DIGIT MINUS SAME DIGIT

Consider the last part of that ditty-rhyme that tells students that when the "numbers the same then zeros the game." First, memorizing and mimicking that rule deprives students of the opportunity to deal with a *number* subtract that *number*, 5 – 5, 2 – 2, 76 – 76, etc.

But then how does that apply to *any other* situation? In 4 + 4, the numbers are the same. In 5 × 5, the numbers are the same. In 3x – 3y, the numbers are the same! Yet in none of these situations is the result supposed to be zero. This is an example of fake math, which is all about fake rules that sort of apply in one grade, but not in any other (Karp et al., 2014).

STEP 3: WHEN THE STUDENT MOVES ONTO THE NEXT COLUMN, SOME OF THE SAME THINGS HAPPEN AGAIN

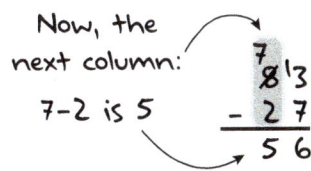

STEP 3, TRAP 1: STUCK IN COUNTING STRATEGIES OR TEENS MINUS SINGLE-DIGIT STRATEGIES, AGAIN

- Count 3 Times
- Count Back by ones

Again, these students are not building or using Additive Reasoning, just getting answers counting one by one.

Other students may be using small number Additive Reasoning, but stuck in this algorithm, they will not grapple with subtraction of larger numbers.

STEP 3, TRAP 2: BEING STUCK IN A DIGIT-ORIENTED APPROACH TO THE ANSWER

Once students have done the last column subtraction, with the answer of 5:

- I got 5, put it next to the 6, so the answer is "5" "6", not acknowledging that 5 in the tens place is 50. The answer is just those digits 5 and 6 next to each other. This digit approach traps students into never really acknowledging the number 56.

- I got 5, put it next to the 6, and 5 next to 6 means 56, doing the steps with digits, but then coming back to acknowledge place value. This can almost be a "do the thing, then think about it" approach. Better, but not having the opportunity to grapple with 50 and 6 being 56.

- 70 − 20 is 50, but there's that 6 already there, so 50 + 6 is 56, thinking about and using place value. This is the best-case scenario in this step of the algorithm, but if this is all the student has been encouraged/taught/given the opportunity to do, they'll just keep subtracting larger and larger numbers, but only ever thinking about *multiples of 10* minus *multiples of 10*.

It "works"! The student using the algorithm got the correct answer to the problem. If they are using a Counting Strategy, they are practicing counting. Even in the best-case scenario, they are at most using teens minus single-digit Additive Reasoning. It is not sufficient for developing Additive Reasoning with subtraction or moving on to Multiplicative Reasoning.

TRAP 5: BEING STUCK IN THE REMOVAL MEANING OF SUBTRACTION

If students are stuck in thinking that subtraction is defined as a *do this algorithm*, a series of steps to memorize and mimic, and one can solve all whole number and decimal subtraction problems using that one procedure, then we are set. Even though there might be a different way of thinking about subtraction, why would you ever want two algorithms to memorize? Students have a difficult enough time with one of them, so don't give them another one. Then, when subtraction as difference/distance and removal comes up in word problems and higher math, it's foreign.

Subtraction has two distinct meanings: removal and distance/difference. The traditional algorithms for subtraction keep learners thinking about removing or minus-ing. In this column, this digit minus that digit, this other digit remove that digit, keep minus-ing/removing in each column until you are done.

But what about the other meaning of subtraction? Consider a basketball game and the score at halftime is 53 − 48. How much do you need to score to catch up? You could find that by removing 48 from 53, but it makes much more sense and is more efficient to just *score up*, finding the distance/difference between 48 and 53.

For 53−48:   You could remove 48 from 53.   Or you could simply find the distance between 48 and 53.

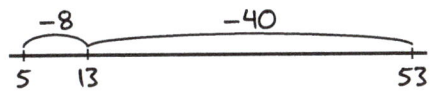

This distance/difference meaning of subtraction happens in comparison problem types. For example,

The dog has 5 toys and the cat has 8 toys. How many more toys does the cat have?

If students have ever only solved subtraction problems using the traditional algorithm, it's counterintuitive for them to consider this a subtraction problem.

But the algorithm keeps learners stuck in removing: For 53 − 48: 3 minus 8, nope; so make it 13 minus 8; now think 4 minus 4. It's always about removing/minus-ing.

This stuck-in-the-removal meaning becomes evident when students think only about removing digits from digits in problems like:

203 − 196; 1025 − 950; or 4.02 − 3.9; or even $6\frac{3}{8} - 5\frac{7}{8}$.

REMOVING DIGITS FROM TEENS OVER AND OVER AGAIN . . .	OR JUST FIND THE DISTANCE/DIFFERENCE
203−196 $\begin{array}{r} {}^{1}\cancel{2}\,{}^{9}\cancel{0}\,{}^{1}3 \\ -\ 1\ 9\ 6 \\ \hline 0\ 0\ 7 \end{array}$ (James Bond?) $_{o^{o^o}}$	$\underset{196 \quad\quad 200\ \ 203}{\overparen{\qquad}\ \overparen{\ }}$ 4 + 3 = 7
1025−950 $\begin{array}{r} 9\,{}^{9}\cancel{0}\,{}^{1}2\ 5 \\ -\ 9\ 5\ 0 \\ \hline 0\ 0\ 7\ 5 \end{array}$	50 + 25 = 75 $\underset{950 \qquad\qquad 1000\ \ 1025}{\overparen{\qquad}\ \overparen{\ }}$
4.02−3.9 $\begin{array}{r} 3\cancel{4}\ \cancel{!}0\ 2 \\ -\ 3\ .9\ 0 \\ \hline 0\ .1\ 2 \end{array}$	0.1 + 0.02 = 0.12 $\underset{3.9 \qquad 4\ \ 4.02}{\overparen{\qquad}}$
A LOT OF FRACTION CALCULATIONS . . .	
$6\frac{3}{8}-5\frac{7}{8}$ $\begin{array}{c} 6\frac{3}{8} = 5\frac{11}{8} \\ -5\frac{7}{8} = 5\frac{7}{8} \\ \hline \frac{4}{8} \end{array}$ $\begin{array}{c} 6\frac{3}{8} = \frac{51}{8} \\ -5\frac{7}{8} = \frac{47}{8} \\ \hline \frac{4}{8} \end{array}$	$\frac{1}{8} + \frac{3}{8} = \frac{1}{2}$ $\underset{5\frac{7}{8} \quad 6 \qquad 6\frac{3}{8}}{\overparen{\quad}\ \overparen{\quad}}$

A few years ago in a workshop where we built the two meanings of subtraction, one middle school teacher pushed back on the idea that the algorithm required a removal/minus approach. "In the algorithm," she said, "I just always found the difference in each column. I wasn't thinking minus/remove. I was thinking distance." Okay, but now she's stuck in distance. We don't actually want students stuck in *always* finding the distance, either. We want students empowered to choose. For problems like 902 – 5, instead of lining those up and doing all the regrouping and thinking what's the difference, just remove 5 from 902. Remember, the purpose of math class is to help your brain develop. That means we help you develop both meanings of subtraction and the intuition for when to use which. And then you will need both meanings to reason about integer subtraction and many higher math topics.

TRY IT

What could it look like to reason additively about 83 – 27?

- Think about 83 – 3. How could you use 83 – 3 to think about 83 – 27?

If 83–3 is 80, how can you subtract the remaining 24?

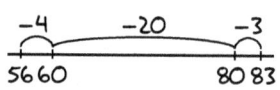

- Think about 83 – 20. How could you use 83 – 20 to think about 83 – 27?

If 83–20 is 63, how can you subtract the remaining 7?

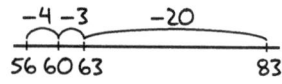

- Think about 83 – 30. How could you use 83 – 30 to think about 83 – 27?

If 83–30 is 53, how can you adjust?

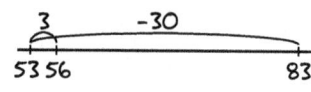

- Think about the distance between 27 and 83. How could you find that distance?

How can you find the distance between 27 and 83?

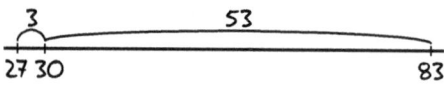

ADDITION AND SUBTRACTION ALGORITHMS WITH BIGGER NUMBERS AND DECIMALS

When numbers get bigger, there are now more columns and the same traps appear when forcing algorithms.

When we introduce decimals into the mix, do we want students to rote-memorize to just move the decimal? Or wait, is this the rule where we line up the decimals? When solving addition and subtraction problems with decimals, the addition and subtraction algorithms require that students line up the decimals and then proceed to attack the columns. Therefore, students have not grown or developed their sense of these different numbers at all. They approach them with, "Line them up and do the same dance as before." No new understanding or sense of magnitudes required. This also means no new understanding or sense of magnitudes is developed.

Because students are not grappling with and developing a sense of magnitudes, a lineup error can occur. Students are doing every step correctly except the first one—they are just lining the number up wrong. Problems with numbers of different digit lengths, like in the following example, are *almost* correct. *Almost,* because every step is correct except the first.

Lining up numbers incorrectly is just one little error.

5194	42.71	348	967
$+\ 36$	$+\ 2.5$	$-\ 21$	$-\ 0.35$
8794	6.771	138	9.32

But it leaves answers that are completely unreasonable.

Students never grapple with larger and more complex numbers in addition. They are getting correct answers, but not building their brain's capacity to reason about the place value and magnitude of the numbers, nor to reason additively about those numbers. These two things play together. Build one, build the other. Repeat.

Conclusion

How important is it really that students can think additively? Isn't it good enough that they can get correct answers? It's harder and more complicated to help students grapple with bigger numbers and the mathematical relationships between those numbers. Couldn't we just memorize this part and then get students thinking when they get to higher math?

A few years ago, I was on the writing team for *Bridges in Mathematics*, a K–5 textbook published by the Math Learning Center. We had written a chapter introduction that described the major models and strategies for multiplication. As I was teaching my university math methods course, I wanted to give that introduction to my students as one of their assigned readings.

Imagine my surprise when I found justification for the multiplication algorithm that had been added.

I emailed the editors. When I pushed back against the statement that "algorithms are more efficient for most (complicated) problems and therefore we need to teach students to use them to solve problems," the editor used 379 × 87 as an example of a problem for which students would "need" an algorithm.

Ignoring for the moment that even if a student did need an algorithm to solve 379 × 87, that wouldn't help the algorithm's abysmal efficiency in learning multiplication. Students absolutely can solve 379 × 87 without an algorithm—especially if they have a solid foundation in Additive Reasoning.

In response, I provided three different approaches a student could take to solve the problem. We'll look at those in Chapter 4. For now, consider that the editor rejected those approaches, not because of the multiplication involved, but because she didn't believe the *subtraction* involved could be done without an algorithm. She stated that 33,060 – 87 was too complex to solve by thinking and reasoning.

This is completely reasonable for someone who had only ever had the opportunity to learn subtraction by using an algorithm. She couldn't fathom students making sense of the numbers. I'll say it again: Algorithms are terrible teaching tools.

33,060-87 with over 10 steps, using a traditional algorithm

$$
\begin{array}{r}
2\ 9\ 15 \\
3\ \cancel{3}\ \cancel{0}\ \cancel{6}\ 0 \\
-\qquad 8\ 7 \\
\hline
3\ 2,\ 9\ 7\ 3
\end{array}
$$

It's a completely different story for students who have been mentored to think like mathematicians. It takes work to develop Additive Reasoning, but when developed, 33,060 – 87 is nearly trivial.

33,060-87 with 2 reasonable, figure-out-able jumps

$$-27 \qquad\qquad -60$$
$$32,973 \quad 33,000 \qquad\qquad 33,060$$

Let's mentor students to reason additively. Because we can. And they can. It's worth it.

Discussion Questions

1. How did you think about addition and subtraction as a student? How has your experience influenced the way you have taught?

2. For these problems: 15 + 9, 99 + 47, 36 – 19, 256 – 99

   a. How do you think about them?

   b. Predict how your students are thinking about these problems.

   c. Ask students an appropriate problem or two. How did they respond?

   d. How might you make your or your students' thinking visible?

3. Many teachers report that they first understood carrying/borrowing the second year they taught it. This underlines how complicated it is. What is your experience? Do you think your experience is common?

4. How are your students reasoning about addition and subtraction situations/problems, using Counting Strategies or Additive Reasoning? To help you decide, you could use the "How are you thinking about addition right now?" questions on page 56 or the "How are you thinking about subtraction right now?" questions on page 79.

5. What are the two meanings of subtraction? Where do those two meanings show up in the grade/content you teach?

6. How has your Additive Reasoning changed while reading this chapter?

7. How have your ideas about teaching Additive Reasoning changed while reading this chapter?

TRY IT IN YOUR CLASSROOM

I Have, You Need

In our base ten number system, the number 10 and powers of 10 are important. By extension, the partners of 10 and partners of powers of 10 are important. For example, 8 and 2 are partners that add to 10; 87 and 13 are partners that add to 100; 788 and 212 are partners that add to 1000. Having these relationships at your fingertips is extremely helpful in solving addition and subtraction problems efficiently with sophisticated reasoning.

For example, $8 + 5$ can be thought of as $8 + (2 + 3) = (8 + 2) + 3 = 10 + 3$, using the partner of $8 + 2$ to get to 10 and then adding the rest.

Similarly, $788 + 356$ can be found by using $788 + (212 + 142) = (788 + 212) + 142 = 1000 + 142$.

To subtract $36 - 17$, you can find an equivalent problem by shifting the distance between 17 and 36 to 20 and 39, so $36 - 17 = 39 - 20 = 19$. And this is all because you recognized that $17 + 3 = 20$.

To help your students (and yourself) build these partners (and more), play the routine we call "I Have, You Need."

I Have, You Need

Purpose

Develop important number partners.

Routine

- Establish a target total.

- Say, "For a total of 100, if I have 88, you need . . . "

- Give brief think time. Cue students to respond with the partner.

- Play a few rounds at a time, gradually increasing the difficulty.

- Alternate between response types: choral, popcorn, down the line, partner play. Keep it moving, keep students anticipating with positive energy. After playing a few times, talk strategy.

Important to Consider

This is not a rote-memorization exercise. It's about having students' brains travel the path of figuring the partners. Doing this often means the path becomes well traveled and familiar. It sets the stage that these partners are interesting and important. Students gain intuition.

Sequencing. Start with numbers closer to the target. For example, when your target is 100, start with decades close to 100, like 80, 60, 70, 90. Then switch to multiples of 5, still close to 100, like 75, 95, 85, 65. Then switch to any number closer to 100, like 88, 92, 76. Then use numbers farther from 100.

Extensions

For young learners, use 10 fingers, ten frame cards, Number Racks with the target total of 5, 10, or 20. For more advanced learners, play with the target total of 60 minutes; the target total of 1 with decimals and fractions; the target total of 90°, 180°, or 360°; the target total of 2π.

Credit and thanks to Kim Montague for creating "I Have, You Need."

CHAPTER 4

The Trap of Multiplication and Division Algorithms

FIGURE 4.1 ● Multiplicative Reasoning is the next domain of reasoning, which includes reasoning about multiplication and division.

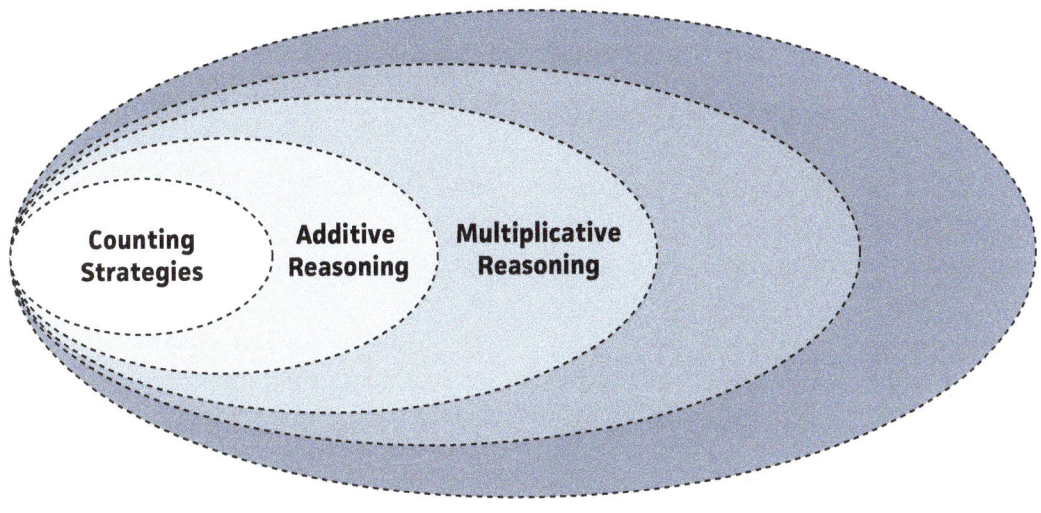

Source: Adapted from Math Is Figure-Out-Able at https://www.mathisfigureoutable.com/ with CC Attribution-NoDerivatives 4.0 International License.

What is 2400 ÷ 24?" asks Kim Montague, my cohost on the *Math Is Figure-Out-Able* podcast. As a guest teacher in this fourth-grade class, she opens the Problem String, reminding students they are to signal with a thumbs-up when they are ready to share their thinking.

As a low-floor access point to the String, most of the class signals they are ready to proceed fairly quickly.

Kim calls on Sawyer, who gives the answer of 100 with the reasoning that when multiplying by 100, the answer will be the original number "bumped up by two place values."

Kim represents his thinking on a ratio table, careful to emphasize the scaling of 1 to 100 and 24 to 2400 as "×100."

Moving on, Kim introduces the slightly less intuitive problem of 1200 ÷ 24. With the helper problem of 2400 ÷ 24 already on the ratio table, the students readily supply that the answer is 50, because 1200 is half of 2400, so half of 100 is 50.

From there, Kim gives the next problem: 1800 ÷ 24.

There is a longer pause on this one. In the thoughtful silence, Kim encourages the students: "You're welcome to write something down. You don't have to hold all your thinking in your head."

TIP

When students are reasoning, using what they know but trying to hold it all in their heads, encourage them to keep track of their thinking by writing it down.

Out of the silence comes two answers from the class: 75 and 110. Obviously, at least one of these is wrong, but Kim betrays no hint, in her words or her body language. In maintaining neutrality, she reinforces that in her math class resilience and learning are valued, not only correct-answer-getting. The correct answer can be highlighted later; there is no getting back the thinking happening right now if it gets shut down.

Callum and Jackson both share their thinking.

The class settles on the answer of 75, justifying that as 1800 is halfway between 2400 and 1200, the answer must be halfway between 50 and 100. This settling is not quick, but Kim doesn't dive in to rescue the confused students. She makes space for them to struggle and, crucially, doesn't leave them there.

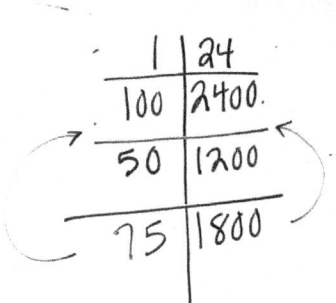

The next problem in the String is 1848 ÷ 24. About 10 seconds pass before Kim calls on Chase, who gives the answer of 77. She then asks if anyone got a different answer, still not betraying through timbre of voice or expression if 77 is correct.

Chase explains that 48 is just two groups of 24 and 1848 is 48 from 1800; therefore, the answer is 2 more than 75 (from the previous problem), so 77.

Next problem: 1776 ÷ 24. As Kim begins to circulate to see what students are thinking, she says, "I wonder if there's anything up there [on the board] that could help you. . . ."

She stops at Avery's desk and asks what she's thinking. Avery notes that 1776 is 24 away from 1800. Kim says, "Hmm, it's only one group of 24 away—that is interesting," and moves on.

Returning to the front of the room, Kim wonders, "Who have I not heard from?" She then calls on Weston, who gives the answer 23, justifying it based on the experience playing "Close to 100." He then changes his answer to 24, but appears increasingly puzzled.

Kim asks, "Do you want some time, or do you want some help?" This is a brilliant teacher move—giving the student agency to choose.

Weston responds that he wants time, so Kim calls on Avery.

Avery proposes using the previously solved $1800 \div 24 = 75$ because 1776 is only 24 away from 1800 so you can just subtract 1 from 75, resulting in $1776 \div 24 = 74$.

Kim broadens the discussion to the whole class at that point, asking generally if anyone else was thinking this way, and connecting it to Weston's early response of 24.

Asking if the class is ready for "one more," Kim presents $2412 \div 24$ and circulates, kneeling down to chat with Brady.

TRY IT

When students share, they suggest three answers: 106, 100.5, and 100 remainder 12. Before you read on, solve the problem yourself. Which answer is correct?

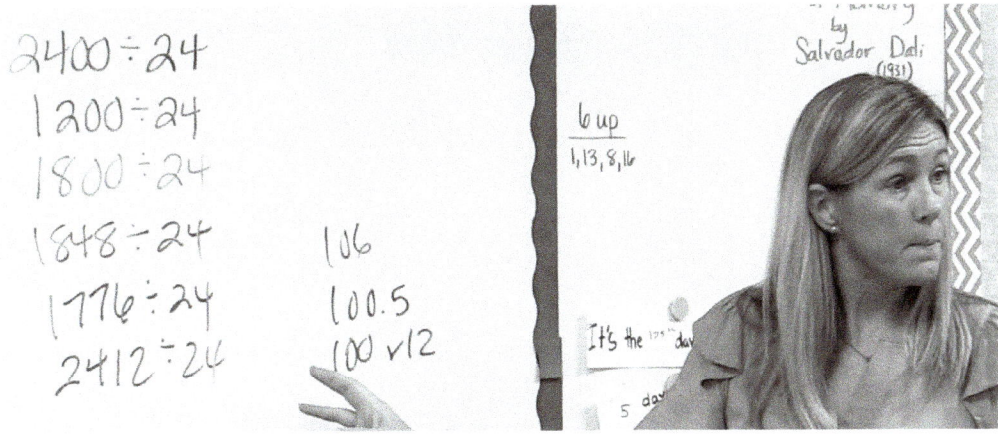

Three students—Aster, Brady, and Caleb—get in on the conversation.

Aster suggests using the 2400 ÷ 24 = 100 and Caleb says, "You can't pack a whole bag, so it won't be 101." Kim adds the 101 to the ratio table to help students reason about one more whole group of 24, 101 × 24 = 2424.

Caleb continues, "So, you could say 100 remainder 12, but it's also half. 12 is half of 24. So, you could also say 100.5."

And Brady agrees, "At the beginning we did 100 times 24 was 2400 and 12 is half of 24."

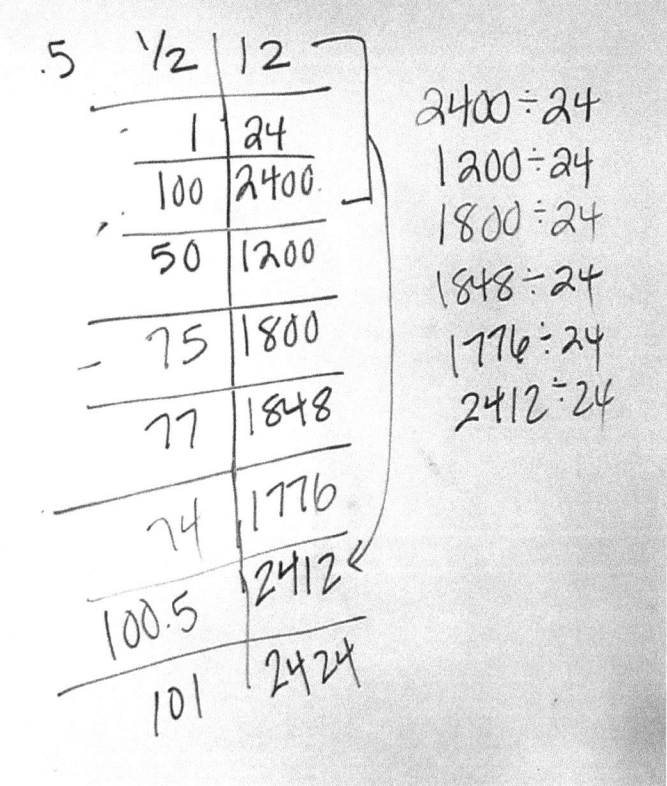

To see these students reasoning in action, watch this video:

https://qrs.ly/x6g3rgk

Kim ends the String by noting that students just did some pretty "gnarly" division problems—decimal division, even—using what they know.

HOW ARE YOU THINKING ABOUT MULTIPLICATION RIGHT NOW?

Was your multiplication instruction primarily algorithms? How do you think about multiplication problems now?

In this section, focus on the way your brain handles these problems. Solve the problem, then read the descriptions underneath, and choose one that best fits your thinking.

How do *you* think about: 38×9 or 98×27?

38×9

- You lined up the numbers and thought: 9×8 is 8, 16, 24, 32, 40, 48, 56, 64, 72. Write down the 2, put the 7 on top of the 3. Then, 9×3 is 9, 18, 27 and $27 + 7$ (from before) is 34. Write the 34 next to the earlier 2 for 342. If you thought about the 9×8 and 9×3 by skip-counting, you were using an additive strategy.

- You lined up the numbers and thought: 9×8 is like 8×8 with one more 8, so $64 + 8 = 72$. Then, for 9×3, you thought that ten 3s is 30, so one less 3 is 27. If you thought about those multiplications using facts you know to reason about them in chunks, you were using single-digit multiplicative strategies.

- If you thought about the problem in big chunks, like 30×9 is 270 and 8×9 is 72 and $270 + 72$ is 342, you were using a multiplicative strategy.

- If you thought about $40 \times 9 = 360$ and subtracted the extra $2 \times 9 = 18$ for $360 - 18 = 342$, you were using a multiplicative strategy.

> **TIP**
>
> Skip-counting means adding one group at a time. Skip-counting is Additive Reasoning, if addition is actually taking place. If students are repeating a song for all the multiples of a number, that is not Additive Reasoning. It's not mathematical reasoning at all; it's singing a song.

98×27

- If you lined up the numbers and did 7×8, 7×9, 2×8, 2×9 each by skip-counting, you were using an additive strategy.

- If you lined up the numbers and did 7×8, 7×9, 2×8, 2×9 each using facts you know (like to find 7×8 by adding one more 7 to $7 \times 7 = 49$, $7 + 49 = 56$ or 7×9 by thinking about 7×10 and subtracting the extra 7, $70 - 7 = 63$), you were using single-digit multiplicative strategies.

- You thought that 98×27 is just a bit less than $100 \times 27 = 2700$. If you thought, "It's just two groups of 27 too big, so $2700 - 54 = 2646$," you were using a multiplicative strategy.

WHAT IS ADDITIVE REASONING IN MULTIPLICATION?

Additive Reasoning in multiplication means finding the total by repeatedly adding the groups, one at a time. For example, finding 6×7 means thinking about six groups of 7 and adding those 7s, one group at a time: 7, 14, 21, 28, 35, 42. It also means keeping track of the number of groups added, stopping when all six groups have been added. It's a multiplicative situation, but Additive Reasoning because you're adding.

The difference between Counting Strategies and Additive Reasoning (counting one by one to dealing with jumps larger than one at a time) is similar to the difference between Additive and Multiplicative Reasoning (adding groups one at a time to grouping the groups, thinking about more than one group at a time).

When students begin learning about multiplication, they must begin considering the number in each group and the number of groups simultaneously (Fosnot & Dolk, 2001).

MULTIPLICATION PROBLEM	BEGINNING STUDENT THOUGHT PROCESS
6 students each brought 8 pencils. How many pencils do we have?	Thinks about 6 students, then about 8 pencils, then thinks about 1 student with those 8 pencils—but really, there are 6 students, each with 8 pencils!

TIP

Here is the juncture where reasoning about the mathematics demands dealing with more things simultaneously than before. Don't skip over the grappling and leave students weak. The instructional routine, Problem Strings, can help give students positive experiences with grappling.

If students have thought about addition solely in columns of single digits, the shift to thinking simultaneously about the number of groups and the number in each group is difficult. They may have been solving problems with bigger and bigger numbers, but because they haven't grappled with those numbers and built brain capacity to grapple with jumps of numbers, they are increasingly reliant on other people's thinking (algorithms) to solve problems. A house built on a foundation of sand cannot last. Thinking about multiplication is *hard enough* when you've already built Additive Reasoning. Without strengthening students' thinking about groups and the amount inside each group at the same time, multiplication is daunting and frustrating. It is why this is the juncture where so many kids decide they are not math people.

When I interact with middle school students, I often ask them 7 × 8, a frequently missed fact, and I hear two consistent responses.

One of the responses is often, "I don't know." This response indicates that students think that 7 × 8 is something you either know or don't know. They believe that 7 × 8 is not figure-out-able. There are numerous ways students can come to this conclusion, but consider how likely such an outlook is if multiplication facts are treated as things to memorize. In a flashcard-driven classroom, figuring out facts is actively discouraged. If we only value speed and accuracy, most students will draw conclusions that answers trump reasoning.

The other response I typically encounter is for the student to take a deep breath and begin, "8 and 8 is 16, then 24, 32, 40. . . ." They skip-count. These students know that 7 × 8 means they need to find seven groups of 8, but they think about it additively, adding each group, one at a time.

> *If students are not adding seven 8s, but rather using a song to sing the multiples or repeating a story or memory technique, that is not reasoning at all. That's rote-memory, which might lead to answers, but it doesn't build reasoning.*

TIP

To students who reply that they do not know a multiplication fact, you could say, "Oh, I didn't ask if you know it. I asked you what it is. What do you know? Figure it out from there!" Send the message loud and clear that the facts are figure-out-able!

TIP

For students who are stuck in Additive Reasoning about single-digit multiplication facts, help them build their Multiplicative Reasoning. Try to build on what they know. Ask questions like, "Are you thinking about 7s or 8s? Do you know some 7s? Some 8s? How many 7s do you know? If you know four 7s, how can that help you with eight 7s? If you know eight 8s, how can that help you find seven 8s?"

WHAT IS MULTIPLICATIVE REASONING?

Multiplicative Reasoning begins with students thinking that repeated additions can be regrouped into bigger groups (van Galen & Fosnot, 2007). As students realize they can group the groups, they chunk the problem into manageable parts. They partition the factors, find the chunks, and then bring those chunks together. Instead of adding each group, one at a time, they add groups of the groups (Figure 4.2).

FIGURE 4.2 ● Counting Strategies, Additive Reasoning, and Multiplicative Reasoning

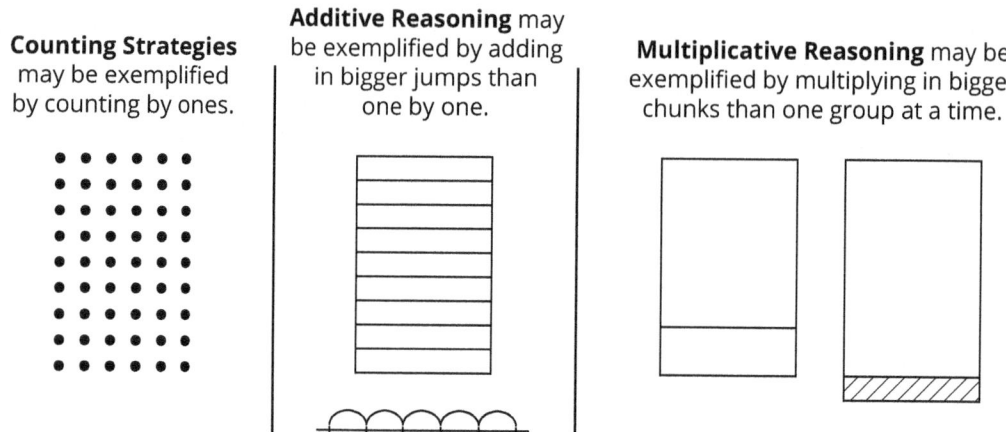

Counting Strategies may be exemplified by counting by ones.

Additive Reasoning may be exemplified by adding in bigger jumps than one by one.

Multiplicative Reasoning may be exemplified by multiplying in bigger chunks than one group at a time.

Source: Adapted from IES (2024). qrs.ly/d8gcfm2

TIP

How can you tell if a student is using Multiplicative Reasoning? If the problem calls for multiplication and the student reasons about chunks bigger than one group at a time, chances are good that the student is using Multiplicative Reasoning. If the student factors one of the factors, the student is using Multiplicative Reasoning.

Multiplicative Reasoning involves reasoning about sets of sets.

"Multiplication involves a new kind of variable, namely the multiplier, which counts *sets*. The multiplier is a property of sets of sets. The multiplicand is a property of sets. . . . Every number refers to sets in addition, whereas in multiplication some refer to sets of sets and others refer to sets. This is a very great difference and the exercise children will have had in dealing with sets and in dealing with sets of sets and even with sets of sets of sets, will serve them in good stead in coming to grips with the problems of multiplication . . ." (Harel & Confery, 1994).

A student demonstrates Multiplicative Reasoning when multiplying 49 × 64 by finding 100 × 64, dividing that in half to get 50 × 64, and then removing the extra group of 64 to get 49 × 64. This can be represented on a ratio table:

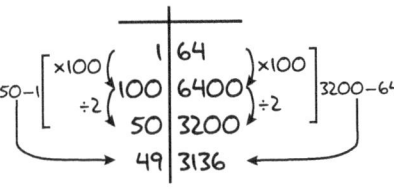

Video of a student solving a multiplication problem using Additive Reasoning:

https://qrs.ly/ zog3rgm

Multiplication can be represented as the area of a rectangle, where the factors are the dimensions of the rectangle and the product is the area.

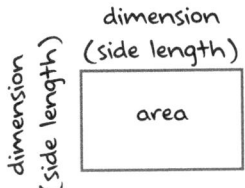

dimension × dimension = area
factor × factor = product

Video of a student solving a multiplication problem using Multiplicative Reasoning:

https://qrs.ly/ zog3rgm

Multiplication can be represented in a ratio table.

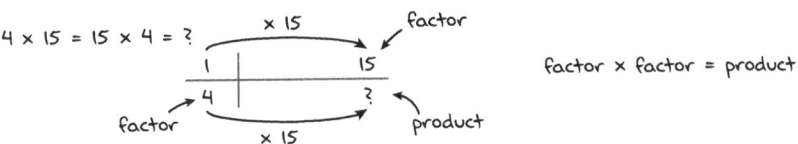

factor × factor = product

Multiplicative Reasoning is also shown in the following examples: 42 × 6.3, 99 × 88, 3.5 × 18, 15 × 86, 75 × 36, 22 × 45.

The six strategies shown in Table 4.1 represent the major relationships and strategies that students need to develop for Multiplicative Reasoning with multiplication.

Video of a student solving a division problem using Additive Reasoning:

https://qrs.ly/ 74g3rgt

TABLE 4.1 ● Major Strategies for Multiplication

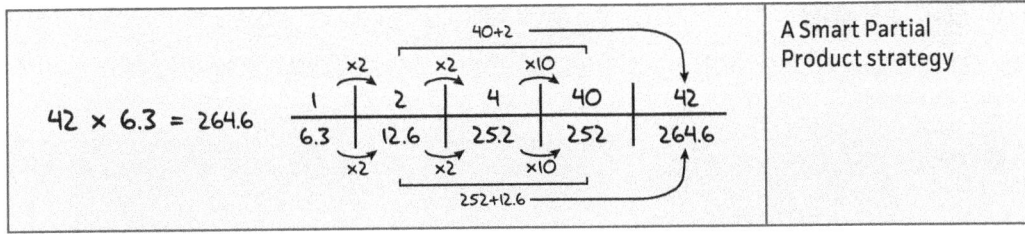

(Continued)

(Continued)

$99 \times 88 = 8712$ 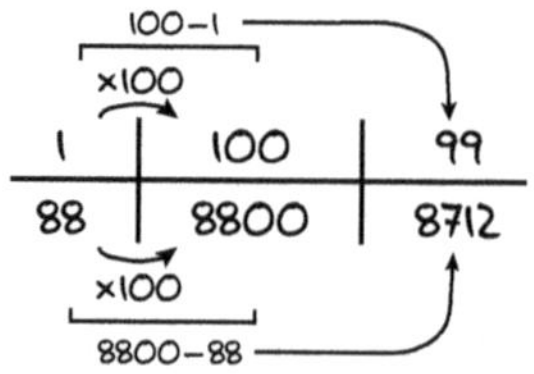 $\begin{array}{c|c|c} & 100-1 \\ \times 100 \\ 1 & 100 & 99 \\ \hline 88 & 8800 & 8712 \\ \times 100 \\ 8800-88 \end{array}$	An Over/Under strategy
3.5×18 $\times 2 \Big(\quad \Big) \div 2$ $7 \times 9 = 63$ 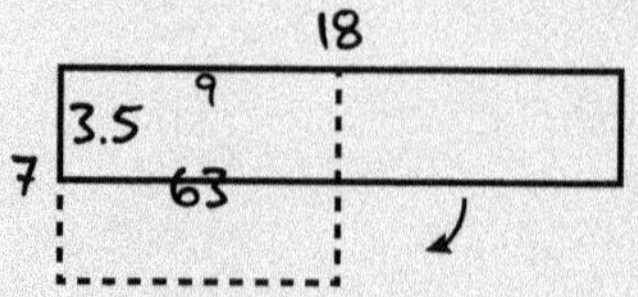 (area model: 18 across top, 3.5 with 9, 7 with 63)	A Doubling/Halving strategy
$15 \times 86 = 1290$ 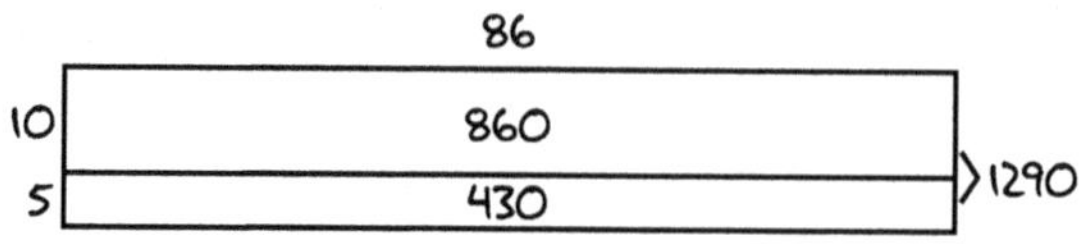 10 → 860, 5 → 430, } 1290 (86 across top)	A Five Is Half of 10 strategy
$75 \times 36 = 3 \times 25 \times 36$ $= 3 \times (100 \times 0.25) \times 36$ $= 3 \times 100 \times (0.25 \times 36)$ $= 3 \times 100 \times 9$ $= 2700$	A Using Quarters strategy
$22 \times 45 = 11 \times 2 \times 5 \times 9$ $= 11 \times (2 \times 5) \times 9$ $= 11 \times 10 \times 9$ $= 99 \times 10$ $= 990$	A Flexible Factoring strategy

Video of a student solving a division problem using Multiplicative Reasoning:

https://qrs.ly/lrg3rgx

With division, a student demonstrates Multiplicative Reasoning when finding $1188 \div 12$ by thinking about $1200 \div 12 = 100$ and then removing the extra group, so $1188 \div 12 = 99$. This can be represented on a ratio table:

$$1188 \div 12 = 99$$

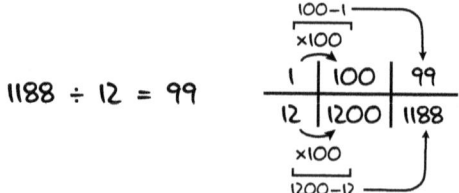

Division can be represented in a ratio table.

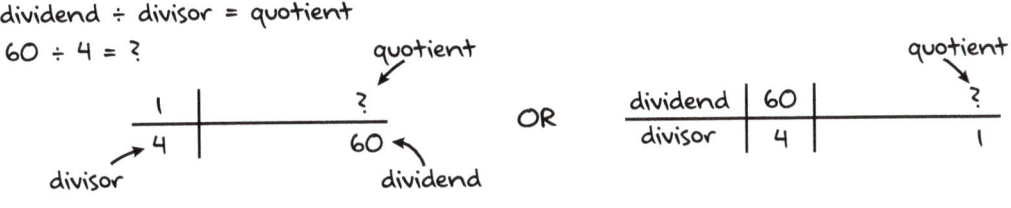

Multiplicative Reasoning with division is modeled in Table 4.2 for $138 \div 6$, $396 \div 22$, $1205 \div 241$, $36 \div 4.5$. The four strategies shown here represent the major relationships and strategies that students need to develop for Multiplicative Reasoning with division.

TABLE 4.2 ● Relationships and Strategies for Multiplicative Reasoning With Division

$138 \div 6 = \dfrac{138}{6} = \dfrac{120}{6} + \dfrac{18}{6} = 20 + 3 = 23$	A Smart Partial Quotient Strategy
$396 \div 22 = 18$ 20−2 ×2 ×10 1 | 2 | 20 | 18 22 | 44 | 440 | 396 ×2 ×10 440−44	An Over/Under Strategy
$1205 \div 241 = 5$ ×10 ÷2 1 | 10 | 5 241 | 2410 | 1205 ×10 ÷2	A 5 Is Half of 10 strategy
$36 \div 4.5 = \dfrac{36}{4.5} = \dfrac{72}{9} = 8$	An Equivalent Ratio strategy

FREQUENTLY ASKED QUESTIONS

Q: Where can I learn more about Multiplicative Reasoning?

A: To learn more about Multiplicative Reasoning and to see it in action, check out the module on Multiplicative Reasoning in my online workshop, *Developing Mathematical Reasoning*, at https://www.mathisfigureoutable.com/dmr/workshop, as well as my upcoming books *Developing Mathematical Reasoning 3–5* (available 2026) and *Developing Mathematical Reasoning 6–8* (available 2026).

The examples in the previous section each show reasoning in bigger chunks of numbers than one group at a time. The traditional algorithms for multiplication and division can trap students into looking like they are multidigit multiplicative reasoners, when they are actually using additive strategies or, better but not good enough, single-digit multiplicative strategies.

Again, if the goal of mathematics education involves learning math and learning to *math*, endlessly practicing skip-counting or rote-memorizing single-digit facts does not support this goal.

THE TRAP OF A TRADITIONAL MULTIPLICATION ALGORITHM

In the United States, textbooks have typically chosen a particular multidigit multiplication algorithm. The steps of that procedure can be summarized as follows: Line up the numbers by place value. Multiply the columns of digits starting with the smallest place value, then regroup as needed. Repeat. Before starting each successive column, record a 0 first, and when the single-digit multiplications are complete, add the rows.

While easy to summarize, this algorithm lays numerous traps for learners developing Multiplicative Reasoning.

Let's walk through the example problem 49 × 24, considering how students could get correct answers while trapped by the algorithm into not reasoning at all about 49, 24, 49 groups of 24, or 24 groups of 49.

Technical multiplication words are factor × factor = product.

STEP 1: FOR 49 × 24, LINE UP THE DIGITS AND WORK WITH THE SMALLEST NUMBERS FIRST

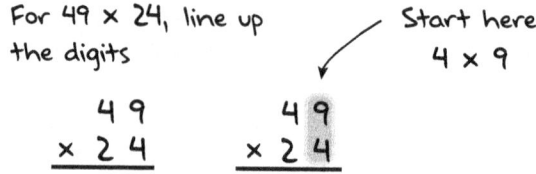

STEP 1, TRAP 1: DIGIT TRAP AND IDENTITY TRAP

Just as with the addition and subtraction algorithms discussed in the previous chapter, this algorithm requires students to treat the numbers as digits—in this case, 4 and 9 and 2 and 4—not as entities 49 and 24. Students are not thinking about *40 something* times *20 something* or even *almost 50* times *a bit more than 20*. They are lining up the digits and then thinking about 4×9, the smallest parts of this problem. When students are not compelled to use these algorithms, their intuition would be to deal with the big numbers first, thinking about 40 and 20, not 9 and 4. Forcing students to consider 4×9 first works against that intuition, sending the message: *Don't think in math class, just mimic. If that doesn't make sense to you, you're not a math person.*

STEP 1, TRAP 2: STUCK IN TRICKS

The order of the steps is difficult for many students to remember, so some well-meaning teachers draw in a turtle to help. This trick might help students remember what to them seem like arbitrary steps to get answers, but it does not help build Multiplicative Reasoning.

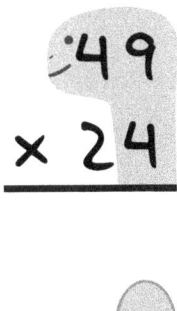

To find 4×9, the student might do one of the following:

- Hold up fingers to do the 9s trick. *A device to get an answer, not reasoning.*
- Sing a song of 9 multiples until getting to the fourth word in the song. *A memory device to get an answer, not reasoning.*
- Repeat a nonmathematical rhyme: 4 times 9 is mighty fine just like dirty sticks is 36. *A memory device to get an answer, not reasoning.*
- Use an external device to retrieve the product like a multiplication chart or calculator. *Literally a device, not reasoning.*

With these memory methods, the student gets the correct answer to 4×9, but is not using Multiplicative Reasoning or really any reasoning at all. They are not building a sense of multiplication. They are not practicing multiplication. It's all about memory.

Of the four, at least there's some math to be found in two of them: the song and the multiplication chart.

While some songs are devoid of any math, the song of multiples is at least numerical. Cue *School House Rock!* from my growing-up years, where PBS cartoons put the multiples to catchy tunes. These songs have some potential to give students a sense of skip-counting. But let's all agree that skip-counting is Additive Reasoning, and so with these songs students are only getting a sense for skip-counting (maybe). If they are recalling the lyrics to a song, they are not actually doing the adding, much less multiplying. For students to progress, we need to work on Multiplicative Reasoning.

The multiplication chart is very mathematical, but in this example it is reduced to a product retrieving tool.

Performing the 9s multiplication trick is not reasoning. Yes, the 9s trick is based on actual mathematics, the fact that 9 is 1 less than 10, which is the base in our number system. But the 9s trick is not transparent and centers around more steps to memorize. When this trick is rote-memorized to provide a cheap way of producing 9s facts, it keeps students stuck, spitting out 9s but not growing their thinking in sophistication. And because it's rote-memorized, what happens if it's memorized wrong? Consider seventh-grade student Michaela, who had rote-memorized the 9s trick, but incorrectly. Because, like so many algorithms, the 9s trick is so opaque, Michaela and other students like her have no recourse to figure out what they are doing wrong. They have to go back to the teacher or online video to re-memorize steps that do not make sense (Figure 4.3).

FIGURE 4.3 ● The 9s Trick

For 4 × 9, put down the fourth finger from the left. The answer is the number of fingers to the left, 3, concatenated (stuck together) with the number of fingers to the right, 6. Therefore, 36.

STEP 1, TRAP 3: STUCK IN ADDITIVE REASONING

Perhaps a student understands that to find 4 × 9, they need four 9s and thinks: I'll add four 9s together, one at a time. 9 + 9 is 18, plus 9 is 27, plus 9 is 36. Once again, this is Additive Reasoning.

STEP 1, TRAP 4: STUCK IN SINGLE-DIGIT MULTIPLICATIVE REASONING

A student trapped in the algorithm could think multiplicatively about the involved single-digits. To find the product of 4 × 9:

- Think 2 × 9 is 18, so just double that to get 4 × 9.

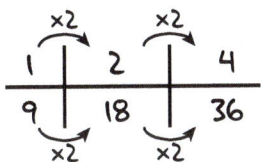

- Think 5 × 4 is 20 and 4 × 4 is 16, so 9 × 4 is 20 + 16 = 36.

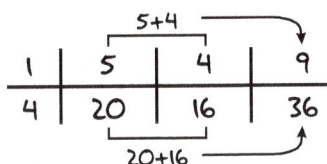

- Think 4 × 10 is 40. That's ten 4s, but since you need only nine 4s, just remove a 4, 40 − 4 is 36.

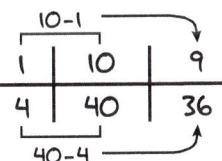

While this is preferable to using no reasoning or using Additive Reasoning alone, students applying single-digit strategies within mimicking steps of the algorithm do not actually grapple with large or more complex numbers. They just keep using single-digit facts over and over. In fact, each two-digit by two-digit problem requires four single-digit multiplications and as many as eight single-digit additions. It's a land of single-digits where students can merrily exist, getting credit for correct answers to larger and larger problems, lulled into a false sense of security. Then these

TIP

Young students beginning to learn multiplication will start by thinking additively. This is a necessary starting point, but it must not be the ending point. The goal is to help students transition from repeated addition to grouping the groups. One way to do this is to ask, "Do you know any bigger chunks?" and "How could you use those bigger chunks?"

"successful" multipliers of large numbers do not recognize completely unreasonable answers when just one mistake throws the entire answer off by a large magnitude. The result is often months spent on place-value units trying to replace the intuitive foundation this algorithm destroyed.

STEP 2: WHAT TO DO WITH THE 36

For 49 × 24, write down the product of that first multiplication, 9 × 4 = 36, by writing the 6 underneath the ones column and the 3 above the tens column.

So 4 x 9 is 36,

Carry the 3 ⟶ 3

Put 6 here

49
x 2 4
6

STEP 2, TRAP: STUCK IN "CARRYING/REGROUPING" USING A DIGIT FOCUS

The student takes the product 36 from the first column multiplication and thinks:

- "The 36 looks like 3 and 6, so put the 6 down there and the 3 up there." They never think about more than digits, and this limited view is devoid of place value. But numbers are more than mere collections of digits. If a student is stuck in this trap, they may very well accidentally write down the 3 in the ones column and the 6 in the tens column because they may forget which number is typically written first. This trap exists with the addition algorithm discussed in Chapter 3. As this is multiplication, the magnitude of the potential error here is multiplicative instead of additive, the direness of which, ironically, can only be appreciated with a developed sense of place value. Instead of adding 60 when they should have added 6, the student is multiplying by 60 instead of 6.

So 4 x 9 is 36. When I write 36,
I write the 3 first, then the 6.

Carry the 6 ⟶ 6

Put 3 here

49
x 2 4
3

- 36 is made of 30 and 6. Put the 3 that stands for 30 up there and the 6 down there because that's what you do, not realizing the place value at work. The student is using place value but focusing on digits, which is slightly more useful. But just slightly.

- 36 is made of 30 and 6. Put the 3 that stands for 30 with the other tens and the 6 with the ones, using place-value understanding. This is the best-case scenario for students inside this algorithm. Some students get this. Most do not.

STEP 3: WHEN THE STUDENT MOVES ONTO THE NEXT COLUMN, MANY OF THE TRAPS REPEAT

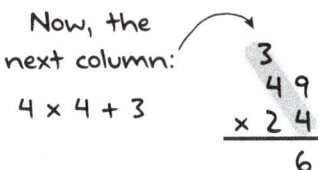

STEP 3, TRAP 1: STUCK IN NONREASONING MEMORY LAND, AGAIN

To find the 4×4:

- Can't do the 9s trick because it's not a 9. Why isn't there a 4s trick?

- Sing the song of multiples of 4

- Repeat a nonmathematical story

- Use an external device to retrieve the product (multiplication chart, calculator)

Again, this student is firmly trapped into retrieving something from rote-memory, not building or using mathematical reasoning.

STEP 3, TRAP 2: STUCK IN ADDITIVE REASONING, AGAIN

To find the 4×4:

The student finds 4×4 by skip-counting 4 and 4 is 8, plus 4 is 12, plus 4 is 16. Practicing addition, not multiplication.

However, there is something even more indicting, er, interesting here. The algorithm calls for the student to find first 4×9 and then 4×4, right? What if they realized the connection between those facts and used this relationship? They could find the 4 × 4 (four 4s) and the 9 × 4 (nine 4s) in the same skip-count! They could only skip-count once and get both products!

Why mention this? After all, we've already established that we don't want to leave students in Additive Reasoning. This is a prime example of what can happen when students are in the trap of thinking that math means "do these steps." In fact, you could do the steps in a different order and save time. How do we get kids to look at the numbers and consider the order? The answer is: Don't trap them in algorithms.

Proponents of using algorithms as teaching tools often cite their efficiency. I would argue there are few things less efficient than doing the same work twice.

STEP 3, TRAP 3: STUCK IN SINGLE-DIGIT MULTIPLICATION

To find the 4 × 4:

Again, while using single-digit multiplication is better than a nonreasoning memory trick or Additive Reasoning, it's not using place value, acknowledging the magnitudes (size) of the numbers. The student finds 4 × 4, but actual place value is behind the scenes—they are using 4 × 4 to compute 4 × 40, but most students do not realize that is what is happening.

STEP 3: MULTIPLY, THEN ADD, THE DIGIT TRAP AGAIN

Once students have multiplied the 4 × 4, they add the regrouped/carried 3.

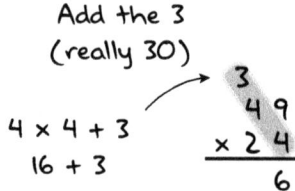

TIP

Easier is not always better. When in doubt, remember that the easiest way to get a student to perform well on a math test is to give them the answer key. Harder is not always better for development, but excessive ease is toxic.

The student is firmly in digit land, finding 4 × 4 + 3. What is really happening is 4 × 40 + 30, but step-mimicking students may never know that. If they've been told, they've surely opted to just think about the digits because that's easier.

STEP 4: WHAT TO DO WITH THE 19

Once the student has found that 4 × 4 + 3 is 19, they write it down next to the 6. Magic! They've done some steps and now have 19 next to 6. Digit focused, that number is rarely, if ever, considered the whole 196.

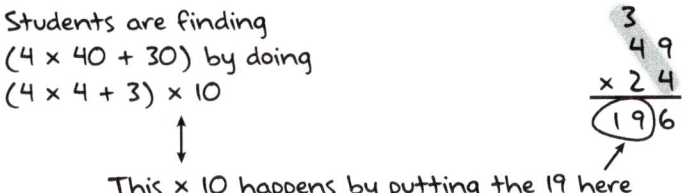

Students are finding
(4 × 40 + 30) by doing
(4 × 4 + 3) × 10
↕
This × 10 happens by putting the 19 here

$$\begin{array}{r} {}^{3}49 \\ \times\ 24 \\ \hline (19)6 \end{array}$$

Why can we just stick that 19 next to the 6? The brilliance of the algorithm's structure is that it lets us record a 19 that is not really 19 at all! It's 190. And 190 + 6 is 196. Another way of thinking about it is that 19 tens + 6 = 196. In the algorithm we see answer-getting without needing to think about the numbers. Students are focused on just doing nontransparent steps correctly. This is not helping brains develop.

STEP 5: THE MAGIC ZERO, THE PLACEHOLDER, THE MASCOT, THE TURTLE EGG?

Now that students are finished with the first row, the next step is to place a zero in the ones place of the next row. This step is so often misunderstood that I have heard teachers say to just put the letter X, a smiley face, or any other nonmathematical notation there because "it's just a placeholder." Remember the turtle? Handy, because now the turtle lays an egg! My four children attended Fuentes Elementary, where their mascot was a star. One of the teachers instructed students to put a star there to "hold the place."

Put a zero here.

$$\begin{array}{r} 3 \\ 49 \\ \times\ 24 \\ \hline 196 \\ 0 \end{array}$$

Or an X.

$$\begin{array}{r} 3 \\ 49 \\ \times\ 24 \\ \hline 196 \\ X \end{array}$$

The turtle lays an egg.

$$\begin{array}{r} 3 \\ 49 \\ \times\ 24 \\ \hline 196 \\ 0 \end{array}$$

Or your school mascot, the stars!
It doesn't matter…
it's just a place holder.

$$\begin{array}{r} 3 \\ 49 \\ \times\ 24 \\ \hline 196 \\ ✶ \end{array}$$

When asked what the zero (or star!) means, many students reply that they have no idea. Granted, it is not important to know if the goal is successfully repeating the steps. But such a goal has dire consequences: Students become more and more convinced that math is not figure-out-able.

And what happens when students forget that magic zero placeholder? Disaster, because with one tiny error, the answer is unreasonable.

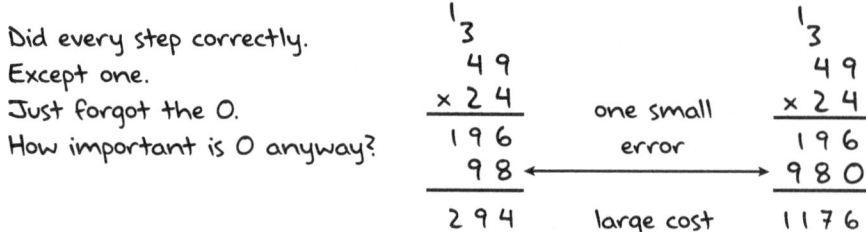

Did every step correctly.
Except one.
Just forgot the 0.
How important is 0 anyway?

one small error

large cost

STEP 6: NEXT SET OF MULTIPLICATIONS

In the next series of multiplications, we encounter the same traps as before, magnified because the digit multiplications are now actually 10 times larger.

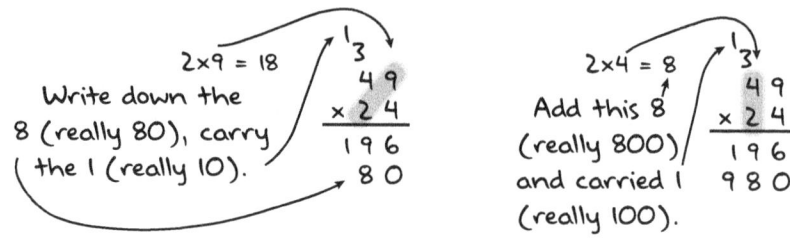

We see:

- Everything is a digit focus.

- Single-digit multiplications, done by memory tricks, additive thinking, or at best multiplicatively thinking about single-digit facts.

STEP 7: ADD THE ROWS

The final row additions at the end bring back *all* of the traps in the addition algorithm, with the accompanying stifling of thinking. Notice that students could be using Counting Strategies in these additions. It's no wonder that some students are so frustrated and exhausted by the time they finish just one problem.

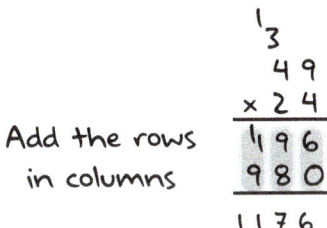

While we're talking about this addition step, let's consider a different problem, 57×27. When using the algorithm, the rows of numbers to add are $399 + 1140$—which means so much regrouping. But if students were thinking, they could simply take $1140 + 400 - 1$. But why even make it that far? The problem 57×27 can be solved by flexibly factoring, $57 \times 27 = 3 \times 19 \times 27 = 81 \times 19 = 81(20 - 1) = 81 \times 2 \times 10 - 81 = 1620 - 81 = 1539$. Flexible factoring is a very multiplicative multiplication strategy.

TRY IT

What could it look like to reason multiplicatively about 49×24?

- Think about 100×24. How could you use 100×24 to find 50×24? How could you use 50×24 to find 49×24?

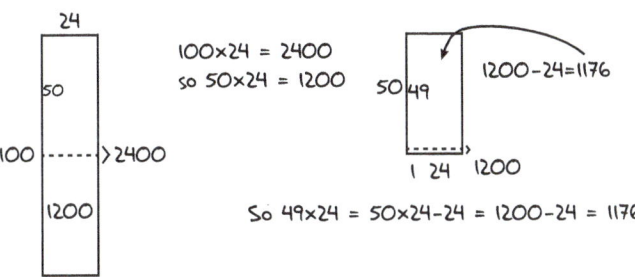

(Continued)

(Continued)

- Think about quarters. How could you use quarters to find 25 × 49? How could you use 25 × 49 to find 24 × 49?

25×49 is one more 49 than 24×49.

Think about 25×49 as 49×25.
So, 49 quarters?

$$49 \times \frac{1}{4} = (48 + 1) \times \frac{1}{4}$$
$$= 12 + \frac{1}{4}$$
$$= 12.25$$
So, 49×25 = 1225

Now 25×49 is 1225.
That's one too many 49s.
So, 24×49 is 1225−49.
1225−49 = 1276−100 = 1176.

- Think about half as many groups that are twice as big. How could you use that to find 98 groups of 12? How could (100 × 12) or (10 × 98) help you find 98 × 12?

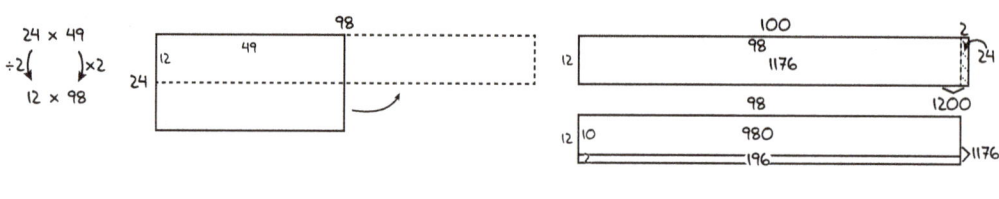

FREQUENTLY ASKED QUESTIONS

Q: Since the traditional algorithm is opaque (all of the place value is hidden behind the scenes), would it be better to do a partial product approach?

A: That is a good first start. Helping students break numbers into their place-value parts and multiply the parts together is the basis of the distributive property of multiplication. Students are dealing with the place values and the magnitudes of the numbers and their multiples. However, leaving students with only a place-value partial product is not sufficient and will be too cumbersome with larger and more complex numbers. Students need more reasoning than just a partial product approach can provide. They need to consider other important relationships that can be learned by developing the major important strategies for multiplication.

Q: What are the major, important strategies for multiplication?

A: There are five important strategies students should learn for multiplication so they can solve any problem that's reasonable to solve without a calculator (Harris, 2022):

1. Smart partial products
2. Over/under
3. Five is half of 10
4. Double/halve
5. Using quarters

There is a sixth strategy that is helpful for many later topics where students need to be thinking even more multiplicatively: flexible factoring, where you factor the factors. These important relationships and strategies help build Multiplicative Reasoning and will help students be in good stead for developing Proportional Reasoning.

An example of each of these strategies is on page 107.

HOW ARE YOU THINKING ABOUT DIVISION RIGHT NOW?

Was your division instruction primarily centered around algorithms? How do you think about division problems now?

The housetop *division notation is made from a right parenthesis (or sometimes a vertical line) and a vinculum (fraction bar). There doesn't seem to be an established formal name, but it is sometimes called the "long division symbol." In this book, we'll refer to it as the housetop division notation (Wolfram Mathworld, 2024).*

In this section, focus on the way your brain handles these problems. Solve the problem, then read the descriptions underneath and choose one that best fits your thinking.

How do *you* think about: 96 ÷ 6, 1152 ÷ 24?

$$96 \div 6$$

- If you think, how many 6s in 96? Then subtract 6 again and again until you get to 0, then count how many 6s you subtracted, you are using an additive strategy.

- You think, this is division and it's not a fact I know, so put the 6 outside the house, the 96 inside. Now, 6 *goes into* 9 once, so write down 1, 1 × 6 is 6, now subtract 9 – 6, that's 3. *Bring down* the other 6, so 36. Ask, does 6 go into 36? Skip-count 6, 12, 18, 24, 30, 36. So, 6 times. If you skip-counted to find the number of times 6 goes into 36, you used an additive strategy.

- If you think, this is division and do the previous steps, but to find the number of times 6 goes into 36, you use a multiplicative relationship like 5 × 6 = 30 so 6 × 6 = 36, so 6 times, you are using a single-digit multiplication strategy.

- You think, what do I know about 6 to multiply up to 96? I know 6 × 10 = 60, so I still need 36. That's six more 6s. So, I have 10 and six groups of 6, so 16 × 6 = 96. Therefore, 96 ÷ 6 = 16. If you thought about chunks of 6 you know to build up to 96, you were using a multiplicative strategy.

- You think, 96 ÷ 6 is like the ratio 96 : 6. Dividing out the common factor of 2 gives you an equivalent ratio 48 : 3 or dividing out the common factor of 3, the equivalent ratio 32 : 2. Either way, they are all equivalent to 16, 96 : 6 = 48 : 3 = 32 : 2. If you were finding equivalent ratios, you were using a multiplicative strategy.

1152 ÷ 24

- You think, this is division, so put the 24 outside the house, the 1152 inside. Now, 24 does not go into 1, and 24 does not go into 11, and 24 *does* go into 115, but how many times? You do a few multiplication problems off to the side, guessing until you find 24 × 4 = 96.

  If you did those side multiplications using skip-counting, you were using an additive strategy. If you did those side multiplications by using single-digit multiplication relationships, you were using single-digit multiplicative strategies.

- You think, what do I know about 24s? How many 24s would get close to 1152? You try 100 × 24 = 2400 and then split that in half to get 50 × 24 = 1200. That's still a bit too high, so you find the difference between 1200 – 1152 = 48. Ah, that's just 2 less 24s than the 50 you had, so you just need 48. If you thought about chunks of 24s like this, you were using a multiplicative strategy.

TIP

Since division is related to ratio and fraction, it is helpful to record a problem like 42 divided by 7 both as 42 ÷ 7 and $\frac{42}{7}$. In general, $a \div b = \frac{a}{b}$.

- You think, that's like the ratio 1152 : 24. Those both have common factors to divide out to make an equivalent ratio. 1152 : 24 = 576 : 12 = 288 : 6 = 144 : 3 = 48. If you were finding equivalent ratios, you were using a multiplicative strategy.

THE TRAP OF A TRADITIONAL DIVISION ALGORITHM

I am cleaning up after dinner when my oldest son, then in seventh grade, asks me a question that stops me in my tracks. For context, this is the kid whose teachers taught all algorithms all the time, but he finds them less useful than what he can come up with on his own.

"Will you please teach me long division?"

Because of the reading and experimenting I had been doing for several years, I have just about made the decision that I was ready to tell the world that we do not need algorithms.

Yet here is my son asking for one.

"Why do you want to learn long division now?" I ask. He'd made it all the way to seventh grade without it! Why now?

"We are doing a ton of problems every day, long division with decimals. I can do all the problems," he says, "but it's taking longer. I'm usually the first done and I'm not anymore."

I glance at his worksheet with *so* many problems of long division of decimals. "You can really reason through all of these?" I will be honest— at the time I had not generalized my ideas to decimal long division.

He nods.

After a brief mental battle—do I hold tight to my growing convictions or give my son a way to get answers quicker?

I grab a piece of paper and write the first problem that comes to mind. I know the dividend should be smaller than the divisor for the answer to be a decimal, $6 \div 8$.

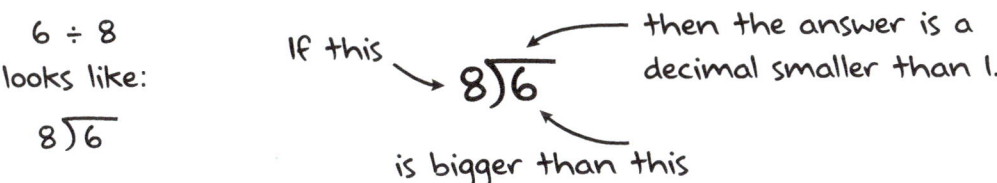

Determined to bring meaning to the steps, I ask, "Okay, kiddo, what times 8 is 6?" as I point to the place above the housetop, the 8 and 6.

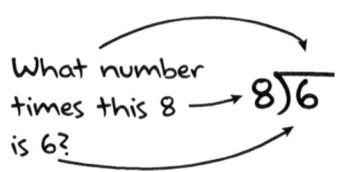

What number times this 8 is 6? → 8)6

I am ready to prompt him with placing the decimal after the 6 and above the housetop, the zero needed to pretend it is 60....

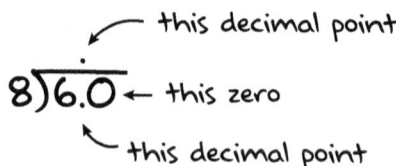

this decimal point

8)6.0 ← this zero

this decimal point

when he looks up at me, thoughtfully, and says, "Um, it's ... two-thirds ... no, it's three-fourths."

Mic drop.

Is he correct? Is $\frac{3}{4}$ of 8 = 6? Sure enough.

$$\begin{array}{r} 0.75 \\ 8\overline{)6.0} \end{array}$$

¾ of 8 = 6
0.75 × 8 = 6

"No," I say to him, my mind blown. "No, I will not teach you long division. Because you have been reasoning through all of these problems, your brain has grown and is reasoning sophisticatedly. Let's not give you a way to stop that progression."

Is division figure-out-able? Yes!

Textbook authors in the United States have typically chosen a particular multidigit division algorithm. The steps of that procedure can be summarized as follows: Put the numbers correctly in and outside the housetop notation. Decide how many times the divisor *goes into* the first digit(s) of the dividend. Write that on top of the house. Write the resulting product under those digits of the dividend. Subtract, regrouping as needed. Bring down the next unused digit from the dividend. Repeat. More succinctly:

- Divide
- Multiply
- Subtract
- Bring down
- Repeat

The long division steps seem so arbitrary and nonsensical (the first step is divide; that's what we're doing—dividing—therefore shouldn't we be finished after this first step?) that teachers have made up mnemonics to help students rote-memorize the steps. Here are a few: Does McDonald's Serve Cheeseburgers? Dad, Mom, Sister, Brother! Do Monkeys Smell Bad? Devious Monkeys Swipe Bananas! If what is being done is focused on making things cute and easy to remember, it's almost definitely treating math as rote-memorizable, not figure-out-able.

This division algorithm is opaque for many students. While I have talked to people who suggest they can think about the place values behind the addition, subtraction, and multiplication algorithms, it is rarer for anyone with the long division algorithm. The place values are at work behind the scenes and this algorithm is so often taught as only procedural that few even try to make sense of it. There are numerous traps for learners developing Multiplicative Reasoning.

Let's walk through the example problem 294 ÷ 6, considering how students could get correct answers while trapped by the algorithm into not reasoning at all about 294, the multiplicative relationship between 6 and 294, or the ratio of 294 to 6.

To begin, the student puts the numbers in the housetop by reversing the order of the numbers from the given problem. The problem began as 294 divided by 6 and now looks like 6 into 294.

Division technical terms: dividend ÷ divisor = quotient.

STEP 1: REVERSE THE ORDER OF THE NUMBERS IN THE PROBLEM

$$294 \div 6 \text{ becomes } 6\overline{)294}$$

It is noteworthy that this is the only time a quotient is written with the divisor before the dividend. Every other representation is written the way you say the problem, 294 divided by 6, but the long division algorithm reverses the order.

For example, 294 ÷ 6, the ratio form 294 : 6, the fraction form $\frac{294}{6}$, and the words "294 divided by 6" are all in the same order, dividend divided by divisor.

STEP 2: GAZINTA

Students ask how many times does the divisor *go into* the first digit of the divisor: "How many times does 6 gazinta 2?"

How many times does 6 go into 2? $6\overline{)294}$

STEP 2: TRAPPED IN A RULE THAT EXPIRES

The answer to "How many times does 6 go into 2?" in the algorithm is "none." Now, I know the question is correctly asked, "How many times does 6 divide *evenly* into 2?" but students all too often internalize that 2 ÷ 6 is a *no-go. Nothing. No answer. Move on to the next step.*

How many times does 6 go into 2? $6\overline{)294}$
None. Move on.

This is a great example of a rule that expires (Karp et al., 2021, p. 22).

This rule, taught in lower grades for a quick fix, expires because you actually can (and do, in later years) divide 2 by 6, 2 ÷ 6 is the fraction $\frac{2}{6} = \frac{1}{3}$.

This is a disaster for students when they are learning fractions and decimals!

Students either:

- roll their eyes at one more time math doesn't make sense;
- compartmentalize that sometimes division means one thing and other times something completely different; or
- decide that since math should be figure-out-able but this time it's clearly not, they are not a math person.

STEP 3: REPEAT, INCLUDE THE NEXT DIGIT

Since 6 did not go into 2, move on and ask the same question for 29.

How many times does 6 go into 29? $6\overline{)294}$

STEP 3: TREATING NUMBERS AS DIGITS TRAP

This may be the first time students have been asked to look at a multidigit number like 294 and consider just the first digit, 2. When considering that the first digit didn't "work," then you consider the first two digits, 29.

You might be thinking that students could consider if 6 divides into 290 by thinking about 29. However, they just had to discount the 2 that represented 200. Now, they can use the 29 that represents 290. One time they could consider place value, but the prior time they shouldn't? For beginning students of division, this is confusing.

STEP 4: WRITE DOWN THE ANSWER

Students find that $6 \times 4 = 24$ and write down the 4 above the 9.

How many times does 6 go into 29?
$6 \times 4 = 24$ so 4 times.

$$6 \overline{)2\ 9\ 4} \quad \text{(4 above the 9)}$$

STEP 4: STUCK IN ADDITIVE REASONING OR SINGLE-DIGIT MULTIPLICATIVE REASONING

To find how many times 6 divides into 29, the student now uses the same strategies they were applying for multiplication: nonreasoning techniques, skip-counting (Additive Reasoning), or single-digit Multiplicative Reasoning. They continue to find 6×4 however they have in the past. They are supposed to be learning about division, but at best they are using single-digit Multiplicative Reasoning.

STEP 4: MULTIPLY?

Students are ironically now told to multiply the newly found 4 times the divisor 6, write that under the 29, and subtract. It's ironic because the first two steps are divide, multiply. That implies that the student divided when finding how many times 6 goes into 29, but now they need to multiply to find that 6 goes into 29 four times and 4 times that 6 is 24. Wouldn't they have already found that when they "divided"? Maybe it's a step because the student has to write it down?

$4 \times 6 = 24$, write that down.
Subtract.

$$
\begin{array}{r}
4 \\
6 \overline{)2\ 9\ 4} \\
-2\ 4 \\
\hline
5
\end{array}
$$

STEP 4: THE TRAPS OF THE TRADITIONAL SUBTRACTION ALGORITHM

When students write the 24 under the 29, the subtraction part of the problem is now written in the same format as in the traditional subtraction algorithm. Therefore, all the subtraction traps are now in play.

It is true that if the students find the correct number to subtract, there will be no regrouping/borrowing necessary, but the rest of the subtraction traps are in full force.

FREQUENTLY ASKED QUESTIONS

Question: Instead of the traditional algorithm, is it better to teach partial quotients, sometimes called the "Lucky 7"?

Answer: A partial quotient strategy is a great starting place, but students must get more efficient than choosing lots of small partial quotients. In the past, teachers have not known what the major, efficient relationships are, but now we do. Also, the "Lucky 7" structure forces students into the subtraction algorithm. So, no, the "Lucky 7" is not a good enough replacement. Instead, help students develop the major multiplication and division strategies. Ratio tables are fantastic models that can be used as tools to reason about multiplication and division but also to solve proportions.

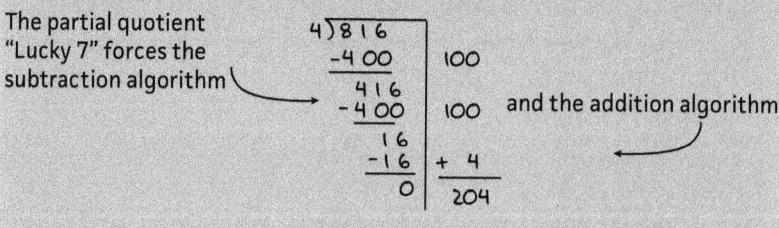

The partial quotient "Lucky 7" forces the subtraction algorithm and the addition algorithm

STEP 5: BRING DOWN

Now, the student *brings down* the next digit in the dividend.

Bring down the next digit.

STEP 5: THE FALSE DEFINITION OF MATH TRAP

In the four-step process, this is the B in the DMBS (*divide, multiply, bring down, subtract*). It's also the least mathematical of the four steps. While the other three steps are operations, there is no such *bring down* operation.

This feeds into the trap that to do math means to mimic repeating steps that make no sense. In this case, *bringing down* is a brand-new thing students have never done before. And it's a *thing to do*, not something to understand. Therefore, in a student's mind, math is not about understanding or reasoning; it's about mimicking and doing.

Many, many teachers have reported to me that when they teach long division, they hear from students, "Just tell me how to do it. What's the next step? No, I don't want to know why or how, just what to do. I'll do the thing and be done with my math work." It gets harder and harder to get students to engage when math is centered around rules and procedures.

STEP 6: REPEAT, WITH ALL OF THE SAME TRAPS AND MORE

The process starts again, where students ask, "How many times does 6 go into 54?" etc. All of the traps are repeated, but now they are worse because the place values are even more obscured.

OVERALL TRAP: BEING STUCK IN THE QUOTATIVE MEANING OF DIVISION

If students are stuck in thinking that division is defined as *do this algorithm*—a series of steps to memorize and mimic—then partitive division in word problems, fractions, and higher math will be completely foreign. The ability to solve all whole number and decimal problems is useless without the ability to interpret context, and long division doesn't just hide context, it actively deconstructs what context a student brought with them by working against their intuition.

The idea of teaching multiple strategies for division in place of a single algorithm when students "can't even get one algorithm" only sounds insane when, like the division itself, the conversation is stripped of the necessary context—that context being that long division is not a helpful way of teaching division. In fact, it's detrimental.

Teachers lament, "I've told students that fractions are division. They just don't get it." These students are stuck in the limited quotative meaning of division from the algorithm.

Division has two distinct meanings: quotative (grouping) and partitive (sharing; Carpenter et al., 2014; Fosnot & Dolk, 2001).

The traditional algorithms for division keep learners thinking about grouping, *How many of these are in that?* This is an important meaning, but we need both.

- Quotative (grouping) division: when you know the number in each group and the total, so you're looking for the number of groups. A quotative strategy can be described as multiplying up from the divisor to the dividend.

- Partitive (sharing) division: when you know the number of groups and the total, so you're looking for the number in each group. A partitive strategy can be described as finding a ratio equivalent to the original ratio.

A robust development of division includes students grappling with both scenarios and using both strategies.

$$1224 \div 24$$

Quotative: How many 24s in 1224? (grouping)

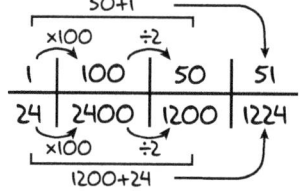

Partitive: Find an equivalent ratio (sharing)

$$\frac{1224}{24} = \frac{612}{12} = \frac{306}{6} = \frac{153}{3} = 51$$

If students are trapped in the long division algorithm, they will crank out step after step for problems that can readily be solved by finding an equivalent ratio.

The equivalent ratio strategy can look like equivalent ratios in a ratio table or like equivalent fractions with equal signs.

DO THE STEPS OF THE ALGORITHM WITH SINGLE-DIGITS OVER AND OVER AGAIN . . .	OR FIND AN EQUIVALENT RATIO THAT'S EASIER TO SOLVE.
Steps of the algorithm with single-digits over and over again . . . $1684 \div 4$ $\begin{array}{r} 421 \\ 4\overline{)1684} \\ -16 \downarrow \\ \hline 08 \\ -\ 8 \downarrow \\ \hline 04 \\ -\ 4 \\ \hline 0 \end{array}$	$\dfrac{1684}{4} = \dfrac{842}{2} = 421$

DO THE STEPS OF THE ALGORITHM WITH SINGLE-DIGITS OVER AND OVER AGAIN . . .		OR FIND AN EQUIVALENT RATIO THAT'S EASIER TO SOLVE.
$3 \div 4$	$\begin{array}{r} 0.75 \\ 4\overline{)3.0} \\ -28 \\ \hline 20 \\ -20 \\ \hline 0 \end{array}$	$\dfrac{3}{4} = 0.75$

In summary, this traditional long division algorithm shares the three main traps:

1. The less sophisticated reasoning trap. Each of the *divide, multiply* steps could be solved using at best single-digit Multiplicative Reasoning. The *subtract* step could be solved using at best single-digit Additive Reasoning.
2. The digit trap. The numbers throughout the process are treated as digits. Students do not have the opportunity to grapple productively with the actual magnitudes involved.
3. The false definition of math trap. Division becomes steps to memorize and repeat like a parrot. (Or maybe like a monkey? Dirty Monkeys Smell Bad, after all.)

Several years ago, in the middle of a workshop for high school teachers, we have been working on dividing numbers, reasoning with ratio tables. I mention that I have seen some inventive ways of helping students rote-memorize the steps of the traditional long division algorithm. The teachers ask me what I mean.

In a flash, I have an idea. I know that the workshop in the adjacent room is *Math for K–5 Teachers*.

"Join me on a quick field trip," I invite and lead the group next door. Luckily, they are just about to take a break.

"If you don't mind, we are working on division, and I am telling these high school teachers that y'all have some ways of helping students. What's your best?"

Those teachers smile, happy to help. They suggest these mnemonic devices to remember (*divide, multiply, subtract, bring down*):

- Does McDonald's Serve Cheeseburgers?
- Dad, Mom, Sister, Brother!
- Do Monkeys Smell Bad?

I thank them. We return to our room, these high school teachers with a clearer picture of why their students are not reasoning multiplicatively about division. They are motivated to dive in and find ways to develop that reasoning in themselves and their students. They are also motivated to relook at their content and the procedures they are helping students memorize. And I am motivated to write this book.

This is not a ding on those elementary teachers. We can only do what we know. It is, rather, an opportunity for us all to realize the trap of these algorithms and choose instead to mathematize with our students.

TRY IT

What could it look like to reason multiplicatively about 294 ÷ 6 ?

- Think about 240 ÷ 6. How could you use 240 ÷ 6 to help you find 294 ÷ 6?

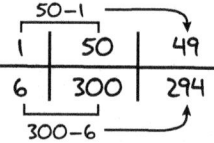

- Think about 300 ÷ 6. How could you use 300 ÷ 6 to help you find 294 ÷ 6?

- Think about 294 ÷ 6 as 294 : 6. How could you find an equivalent problem that is easier to solve?

Division can look two different ways on a ratio table.

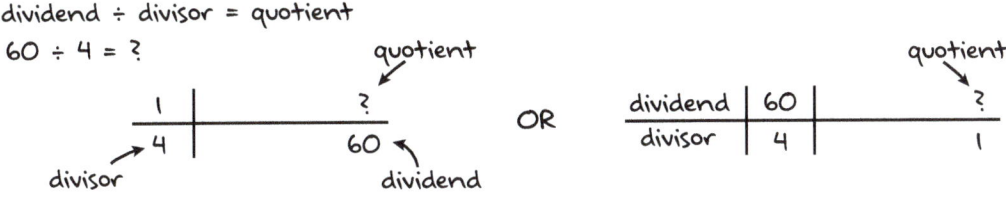

Division can be represented on an open array (rectangle).

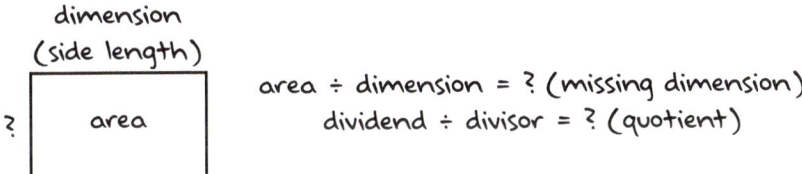

FREQUENTLY ASKED QUESTIONS

Q: What are the major, important strategies for division?

A: There are four major, important strategies that students should learn for division so that they can solve any problem that's reasonable to solve without a calculator: smart partial quotients, over/under, five is half of 10, and equivalent ratio. Examples of each of these are on page 109.

MULTIPLICATION AND DIVISION WITH BIGGER NUMBERS AND DECIMALS

Just like with addition and subtraction, larger magnitudes and decimals compound the traps that algorithms lay in multiplication and division.

The algorithms are brilliant because they are basically the same steps, just with the one additional "move the decimal" requirement. It's amazing that there are procedures where with just one tiny extra step, one could solve problems with bigger and more complex numbers. While algorithms require no new understanding to operate with large magnitudes and decimals, it also means no new understanding of multiplying/dividing with large magnitudes or decimals is developed.

Moving forward with this lack of understanding is less building on a sandy foundation for fractions and Proportional Reasoning, and more building on no foundation at all. It is worse! This turns multiplication and division with decimals into the same dance as before, but with more digits and "magic" moving decimals. Magic decimals are not understood decimals.

Allow me to illustrate the traps of algorithms in decimal multiplication by describing an experience Kim Montague had with her former student, Michaela. Michaela had Kim in fifth grade, with all the reasoning and understanding that implies.

When Kim runs into Michaela's mother two years later, she mentions that Michaela is struggling in seventh grade and they agreed that Kim will tutor her.

At that first tutoring session, Michaela tells Kim she needs help with decimal multiplication. When Kim hears her talking through a problem, she asks, "Will you please do that on the board? Pam's going to want to see this." She was so right!

So, Michaela does the problem on the white board, with Kim videoing. Notice that for the problem 0.26 × 24, Michaela lines up the decimals, adding zeros where necessary. She tells Kim that her teacher says it doesn't matter whether it's addition/subtraction or multiplication/division; line up the decimals every time.

Source: Kim Montague

Then Michaela proceeds to do all of the single-digit multiplications. At the end of the first line, she draws two loops and the decimal while saying, "And then I do the 'butt-cheek method.'"

Source: Kim Montague

After she adds the two lines together, she finishes by looping and drawing the decimal as she says, "Then butt-cheek method.

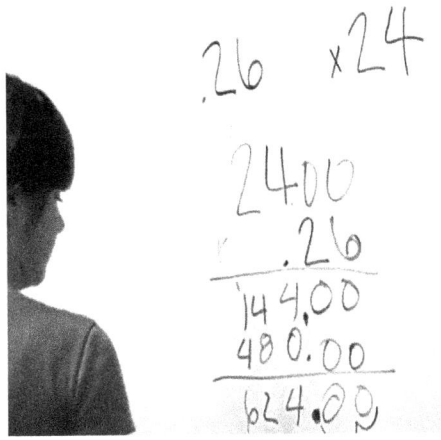

Source: Kim Montague

And the answer is 624 hundredths."

I have shown this video to thousands of teachers, and it never ceases to surprise me when teachers say, "Oh she was almost right. So close!" when in reality, her answer is very far off. They mean, of course, that her only mistake is that she butt-cheeked incorrectly. This notion of being almost right when you've done all but one of the steps correctly is rooted in answer-getting, not in the purpose of math class, which is to build mathematical reasoners.

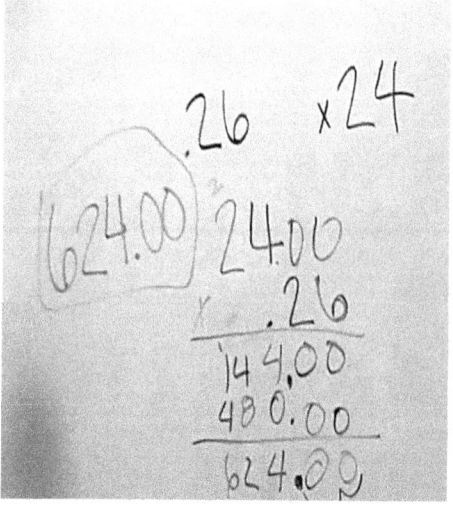

Source: Kim Montague

To watch the butt-cheek video:

https://qrs.ly/cog3rgz

To watch Michaela rock decimal multiplication with the *same* problem:

https://qrs.ly/o1g3rh0

Off-camera, Kim asks Michaela why she only "butt-cheeked" twice. Looking slightly sheepish, Michaela answers, "Ms. Montague, we only have ... two ..." and she trails off, "butt-cheeks."

The only thing worse than a magic-decimal is a magic-butt-cheek-decimal.

When Kim follows by asking, "Michaela, how do you *think* about problems with decimals?" Michaela answers, "Oh, my teacher told us that we need to memorize this stuff," implying this stuff must be memorized—not thought about—because it's not figure-out-able.

Kim works with Michaela for a few 1-hour sessions, refreshing all the reasoning they had done in fifth grade. Pretty soon, Michaela is reasoning about decimals like a champ.

During one of those sessions, Kim gives Michaela the same problem, 0.26×24. To summarize, Michaela takes one look at the problem and says, "Well, 0.25 of 24 is a quarter of 24, which is 6. And 0.01 of 24 is 0.24. So, the answer is 6.24."

Bam.

$$7.2 \div 0.8 \rightarrow 0.8\overline{)7.2} \rightarrow 0.8\overline{)7.2} \rightarrow 8\overline{)72} \quad \text{because}$$

	×10	
7.2	72
0.8	8
	×10	

Conclusion

In Chapter 3, I mentioned an argument—er, discussion—I had with the editors of the *Bridges in Mathematics* program, where they suggested that "algorithms are more efficient for most (complicated) problems and therefore we need to teach students to use them to solve problems," and the example problem they suggested was 379 × 87.

A conservative estimate of the number of steps it takes to find 379 × 87 using the traditional algorithm is 22 steps, and that's if students are using single-digit multiplication strategies. If they are using skip-counting, the number of steps is much higher. Either way, students are stuck doing all of those steps *and* trapped in less sophisticated thinking. In my email response, I suggested the following three strategies.

1	100	10	90	2	3	87			
379	37,900	3,790	34,110	758	1,137	32,973			

37,900 − 3,790 = 38,110 − 4,000 = 34,110

34,110 − 1,137 = 34,973 − 2,000 = 32,973

1	3	300	0.75	75	4	379			
87	261	26,100	65.25	6,525	348	32,973			

1	2	4	400	20	380	379			
87	174	348	34,800	1,740	33,060	32,973			

34,800 − 1,740 = 35,060 − 2,000 = 33,060

33,060 − 87 = 32,973

To be clear, I'm not suggesting when you are in the middle of living your life and have a desperate need to solve 379 × 87 that you use these strategies *or* the algorithm. I think this is a fine case just to use technology. But notice that within the first few steps of any of my three strategies, I've got a useful approximation. And, most of all, I ask you to consider what is the purpose of math class—to get answers to questions like 379 × 87 or to develop a student's brain to reason multiplicatively? Which takes us further, the answer 32,973 or a stronger, more complex thinking brain?

Discussion Questions

1. Are there any single-digit multiplication facts you don't know instantly, that you have to refigure when you need them? How do you figure them: a rhyme or story, skip-counting, some multiplicative strategy? How do you wish your brain was inclined to find them?

2. How did you think about multiplication and division as a student? How has your experience influenced the way that you have taught?

3. For these problems: 15 × 9, 99 × 48, 144 ÷ 16, 288 ÷ 32

   a. How do you think about them?

   b. Predict how your students are thinking about these problems.

   c. Ask students an appropriate problem or two. How did they respond?

   d. How might you make your or your students' thinking visible?

4. How are your students reasoning about multiplication and division situations/problems: using Additive Reasoning, single-digit Multiplicative Reasoning, or more sophisticated Multiplicative Reasoning? To help you decide, you could use the "How are you thinking about multiplication right now?" questions on page 103 or the "How are you thinking about division right now?" questions on page 121.

5. What are the two meanings of division? Where do those two meanings show up in the grade/content you teach?

6. How has your Multiplicative Reasoning changed while reading this chapter?

7. How have your ideas about teaching Multiplicative Reasoning changed while reading this chapter?

TRY IT IN YOUR CLASSROOM

Doubling and Halving

Learning to recognize and work with doubles and halves of numbers is very helpful when developing Multiplicative Reasoning.

Doubling numbers can be a first multiplication strategy for finding times 2, times 4, and times 8. This is true for often-missed facts like 8×7 ($2 \times 2 \times 2 \times 7 = 2 \times 2 \times 14 = 2 \times 28 = 56$), but also for larger and more complex numbers, like $8 \times 213 = 2 \times 2 \times 2 \times 213 = 2 \times 2 \times 426 = 2 \times 852 = 1704$ and $8 \times 4.2 = 2 \times 2 \times 2 \times 4.2 = 2 \times 2 \times 8.4 = 2 \times 16.8 = 33.6$.

Doubling shows up in powers of 2, population (exponential) growth $y = 2^x$, the base two number system upon which computers are based, and computer memory sizes (64, 128, 512, etc.). If a number is a double, then the number is even, it can be halved into two whole numbers, it can be represented by a 2 by something array, it has a factor of 2, and so on.

Halving shows up with doubles inversely, and also with fair sharing, finding equivalent fractions by dividing out the common factor of 2, helps with estimation and reasonableness, and much more.

Put doubling and halving together and you get 1, $2 \times \frac{1}{2} = 1$. You also get a great multiplication strategy, where you double one factor and halve the other to create an equivalent problem that is easier to solve, such as $45 \times 18 = 90 \times 9 = 810$, or even $12.5 \times 2.4 = 25 \times 1.2 = 50 \times 0.6 = 100 \times 0.3 = 30$.

One of the best reasons to double and halve with students is that mathematicians play with doubles and doubling, halves and halving. Let's all math like mathematicians!

Purpose

Develop Doubling and Halving

Routine

- Suggest a starting number.

- Ask students to double the number.

- Repeat.

- When the doubling takes some thought, elicit students' strategies for doubling and make them visible with a model. Compare for efficiency, cleverness.

(Continued)

TIP

Do routines like this often and in short bursts. Write a short list of high-impact instructional routines on an index card and hang it by your classroom door as a reminder. In spare minutes, while waiting for class/the day/math time to end, grab the card and double or halve with students during that time.

(Continued)

- After students have doubled enough, change the routine to halving.

- Keep it to snappy, short bursts; emphasize reasoning.

Important to Consider

Don't make this a speed competition; rather, focus on students using what they know, striving for clever strategies. This is also not a rote-memorization exercise, but having students' brains travel the path of figuring doubles. Doing this often means the path becomes well traveled and familiar. It sets the stage that doubles are interesting and important and helps students gain intuition.

Having students work to double serves well as a differentiated task too. Those who are new to Multiplicative Reasoning can use additive thinking to add the number to itself, while hearing and seeing what others are doing to double.

Sequencing. Aim for a sweet spot for the numbers you choose—not too easy and not too difficult. You can start with an easy number and have students continue to double the doubles—that will get difficult soon enough. Try 12, then 13; and 20, then 19. Change it up by choosing random numbers, and sometimes let students choose the numbers.

Extension

When appropriate, choose numbers to double like 3.5, 35, 350 or $\frac{3}{4}$, 7.5, 750, and compare the place values of these numbers and their doubles. When halving, don't omit odd numbers, like 9 or fractions like $\frac{1}{3}$.

CHAPTER 5

The Trap of Fraction- and Proportion-Solving Algorithms

FIGURE 5.1 ● Proportional Reasoning is the next domain to build in the progression.

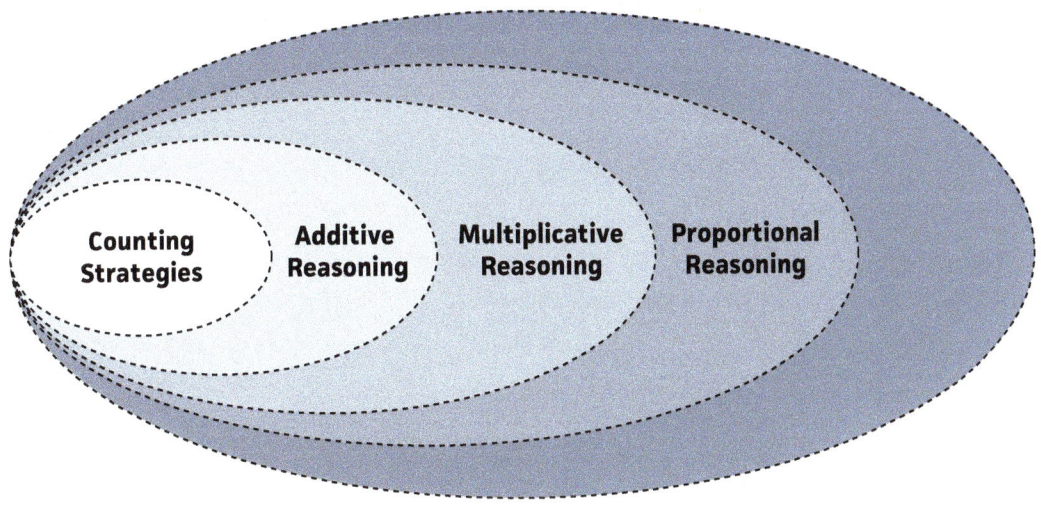

Source: Adapted from Math Is Figure-Out-Able at https://www.mathisfigureoutable.com/ with CC Attribution-NoDerivatives 4.0 International License.

Recently I led a two-day Proportional Reasoning workshop—and, I have to be honest, the teachers in this particular workshop start our time together anxious and a tad hostile. I can't blame them. Unbeknownst to me, the school district has made the training required *and* neglected to tell any of the teachers before they show up that we would be filming. (I end up getting permission from those who would give it and blurring out the few who didn't.)

Understandably, emotions run high. Many of the teachers do not want to be there, and I need to challenge their beliefs about teaching fractions and proportions to get anything done. Rare is the soul who enjoys having their beliefs challenged, even on the best of days. Which makes my interactions with a specific teacher, Melanie, even more striking.

Polite and full of strong emotion, Melanie lays out her argument of how her population of students could only be hurt by the teaching practices I am advocating. *This is too hard—they need the basics and only one way.*

Familiar with this position and after interacting for a bit, I have a hunch that Melanie is not reasoning proportionally and there-fore, understandably, does not believe her students can, either.

In that moment there is no magic reassurance I can give her. She needs more experience. So, I honor her hard work and encour-age her to keep going. The ensuing conversations are neither quick nor comfortable, but perseverance and respect from both of us allow her to keep learning.

The next day, as I facilitate a Problem String about slices of pizza, every-thing changes for Melanie. Her face lights up and her eyes open wide. She has solved a problem using what *she* knows and understands. She has reasoned proportionally and, more importantly, realized that she could have been reasoning proportionally this entire time. That "keep-change flip" or "invert and multiply" was only ever the most fossilized remains of real mathematics (Freudenthal, 1973). Reasoning was always within her reach, only she did not know it was an option.

In that crystallized moment so powerfully provided by personal experience, so directly opposite of algorithm-centered teaching, Melanie knows that not only can her students do what she did, but that they will be happier, stronger, and more powerful for it.

I know this because she tells me, as she will now tell anyone who asks: "Today, I actually understood it. I actually got it for the first time."

To hear from Melanie, check out the QR code on this page.

Watch Melanie share her experience

https://qrs.ly/ rsg3rh5

HOW ARE YOU THINKING ABOUT OPERATIONS OF FRACTIONS RIGHT NOW?

Was your own instruction about fraction operations primarily algorithms? How do you think about solving problems involving fractions now?

In this section, focus on your instincts. How does your brain want to attack the problem? Solve the problem, then read the descriptions underneath, and decide if any match your thinking.

How do *you* think about finding:

$$\frac{45}{60} \times \frac{4}{17}$$

If you think, "This is the rule where you find a common denominator," or "flip and multiply," or "cross multiply and divide," then you are retrieving a rule from rote-memory, but you've misremembered/guessed the wrong rule.

If you think, "This is the rule where you multiply across," you've remembered/guessed the right rule. If you do the multiplications 45×4 and 60×17 using the traditional multiplication algorithm, you could be finding each of the algorithm's single-digit multiplications using rote-memory techniques (not mathematical reasoning), Additive Reasoning (skip-counting), or single-digit multiplicative strategies. If you do those multiplications using multiplicative strategies like double/halve to find $45 \times 4 = 90 \times 2 = 180$, you are using multiplicative strategies.

If you think about $\frac{45}{60}$, consider the relationship of 45 and 60, and realize that $\frac{45}{60}$ is equivalent to $\frac{3}{4}$. Then you might think about $\frac{3}{4} \times \frac{4}{17}$ as $\frac{3}{4}$ of $\frac{4}{17}$. Since $\frac{4}{17}$ is $4 \times \frac{1}{17}$ or 4 one-seventeenths, you can think about $\frac{3}{4}$ of 4 one-seventeenths. Since $\frac{3}{4}$ of 4 things is 3 of those things, then $\frac{3}{4}$ of $\frac{4}{17}$ is 3 of those one-seventeenths, $\frac{3}{17}$. If you thought using these relationships, you were using Proportional Reasoning.

TIP

Do not conflate the verbosity of explaining a strategy with the difficulty of using a strategy. This example of Proportional Reasoning is also an example of when it takes a lot more time to write out the thinking than it takes to think it. In a proportional reasoner's head, solving $\frac{3}{4}$ of $\frac{4}{17}$ quickly translates into solving $\frac{3}{4}$ of 4, which is not step-intensive at all.

HOW ARE YOU THINKING ABOUT SOLVING PROPORTIONS RIGHT NOW?

How do *you* think about finding the unknown in these proportions?

$$\frac{4}{12} = \frac{a}{60} \quad \text{and} \quad \frac{7}{8} = \frac{d}{41.6}$$

Solving for *a* in:

$$\frac{4}{12} = \frac{a}{60}$$

You might remember some version of "cross multiply and divide," so you multiply 4 × 60 and divide that by 12. If you did that multiplication and division using an algorithm, you could have been using counting or Additive Reasoning, repeating rote-memory devices, or at best single-digit multiplication strategies. If you did the multiplication and division using Multiplicative Reasoning, you were using a multiplicative strategy.

You might think that to get from 4 to 12, you would scale times 3. You ask yourself what scaled times 3 is 60? Since 20 × 3 is 60, a must be 20. That would be scaling in tandem *within* each ratio, which is a Proportional Reasoning strategy.

$$\times 3 \left(\dfrac{4}{12} \, \bigg| \, \dfrac{20}{60} \right) \times 3$$

You might think that to get from 12 to 60, you would scale times 5. Therefore, you would scale 4 times 5 to get $a = 20$. That would be scaling in tandem *between* the ratios, which is a Proportional Reasoning strategy.

$$\overset{\times 5}{\underset{\underset{\times 5}{}}{\dfrac{4 \mid 20}{12 \mid 60}}}$$

Solving for d in:

$$\frac{7}{8} = \frac{d}{41.6}$$

You might remember some version of "cross multiply and divide," so you multiply 7 × 41.6 and divide that result by 8. If you did that multiplication and division of decimals using an algorithm, you could have been using counting or Additive Reasoning, repeating rote-memory devices, or at best single-digit multiplication strategies. Since there are decimals involved, you might have employed the butt-cheek strategy (see Chapter 4).

Using a mnemonic means you are treating this mathematics as social knowledge. Handling decimals via memorized tricks instead of a foundation of reasoning will prevent students from developing intuition and will provide a shaky foundation for Proportional Reasoning and beyond (whether the mnemonic involves butt cheeks or not).

You might think about how to scale from 8 to 41.6, knowing that you would also scale 7 by that factor. Since 8 × 5 is 40, you still need to scale from 8 to the leftover 1.6. Since 8 × 2 is 16, 8 × 0.2 must be 1.6. Therefore, you could also scale 7 by 5 and then by 0.2. That would be reasoning proportionally.

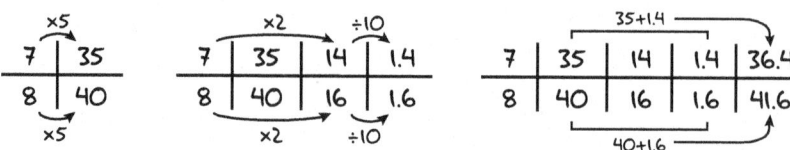

If reasoning through these problems seems daunting, take heart. Proportional Reasoning is figure-out-able. But it also takes intentional work to build. And Proportional Reasoning is multiplicative, so if you're just starting to build your Multiplicative Reasoning, you may be where I was a few years ago.

As I was diving into journals and books early in my journey, someone recommended Susan Lamon's book *Teaching Fractions and Ratios for Understanding*. Remember that at this time, I had been a highly successful math student and was currently a successful high school math teacher.

At the beginning of her book, Lamon suggests that over 90% of adults do not reason proportionally. When I read that, I didn't even pause. Of course, that wasn't me. I wouldn't be in that 90%. She then suggests that if you are reasoning proportionally, you should be able to reason through a set of 10 problems that she lists. The directions said not to set up a proportion and cross multiply and divide—you had to reason your way through.

I thought, "No problem. I am a successful high school math teacher. I got this."

I read the first problem.

I skipped the first problem.

Then I skipped a few more problems.

When I was finished, I could solve five of them using an algorithm, but for the rest I had no idea. For one I could remember learning a formula, but I'd only done that kind of problem a few times so I had forgotten the procedure. In the world of rote-memorizable fake math, that meant I had no recourse for that problem, not to mention the others that I don't think I had ever been asked to solve.

I was stunned. But I was also confused. I didn't really know what I was missing.

Being busy in that time of life, I set that book and its accompanying disequilibrium aside and continued to work with my young children and their teachers, building my Additive Reasoning and then my Multiplicative Reasoning.

Somewhere in that later time when I was getting multiplicatively strong, I picked up Lamon's book again. Whoosh! I had access. My brain was stronger and could grapple with more simultaneously. I was looking at problems as figure-out-able, asking myself what I knew that could help and building off that. I could reason through those problems! I was beginning to reason proportionally.

WHAT IS PROPORTIONAL REASONING?

One of my more recent favorite moments in a middle school class-room happened while I was the guest teacher filming a Problem String designed to build Proportional Reasoning. These students had no prior experience with Problem Strings.

I begin by setting the stage that the new pizza place in our town is selling pizza by the slice. The deal is 4 slices for $5.

First question: If they'll sell it, how much do 2 slices cost? Students easily cut 4 slices in half and find the corresponding price.

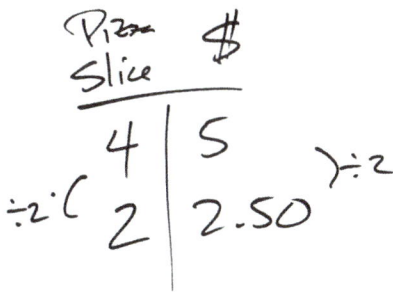

When I ask for the cost of 3 slices, we compare a few strategies. Almost everyone finds the unit price and then either adds the prices for 1 slice to 2 slices (more additive) or scales the price for 1 slice times the 3 slices (more multiplicative) as shown.

Pizza Slice	$
4	5
2	2.50
3	3.75
1	1.25

1+2 [bracket on 2, 3, 1]

Pizza Slice	$
4	5
2	2.50
3	3.75
1	1.25

×3 (bracket on 3, 1) ×3

One student talks about how the cost for 3 slices would be right in the middle of the cost for 2 and 4 slices since 3 was right in the middle of 2 and 4 slices. They average the cost of 2 and 4 slices.

Pizza Slice	$
4	5
2	2.50
3	3.75
1	1.25

3 → ← 3.75

Then I ask how much it would cost for 10 slices. Students share several strategies, including finding 8 slices and adding that to 2 slices, finding 5 slices and doubling, and also scaling from the 1 slice to 10 slices.

Pizza Slice	$
4	5
2	2.50
3	3.75
1	1.25
10	12.50
8	10.
5	6.25

×10 (bracket on 1, 10) ×10

The fun really kicks off when I ask, "If they'd sell it to me, how much could I get if all I had was $1.00?"

Zyahna quickly says, "Just cut off the crust." And her partner, Adri, says, "I was thinking the $0.25 is for the toppings so get a plain pizza for a dollar."

I smile because they are reasoning that we would get less than a whole piece. A nice start to finding a more exact answer.

When I ask students to share their answers, they respond with a variety: 0.75, 4/5, 8/10, 0.8, and 80%. Brandon explains that he subtracted 0.25 from the 1 slice and from the $1.25, resulting in the ratio 0.75 : 1.

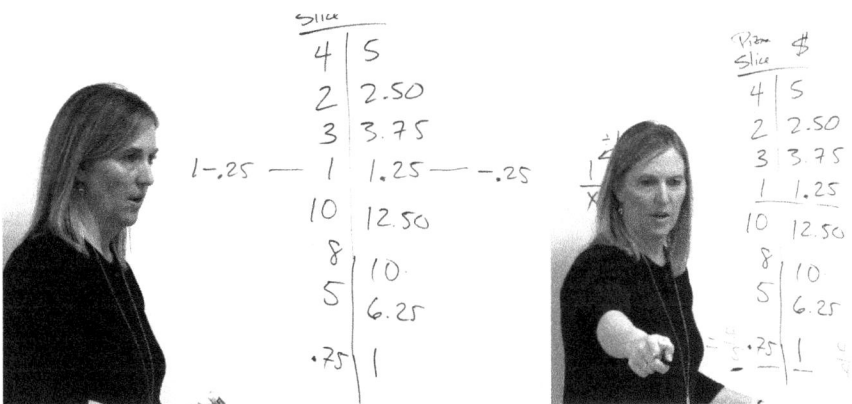

As I note that their answers don't agree, one student suggests, "We could just round up the 0.75," earning a chuckle from everyone. I mean, it's a slice of pizza—0.75 of a slice is pretty close to 0.8 of a slice. Just round up?

After several students share how they had found 0.8, I ask the questions, "Will a quarter of a dollar buy a quarter of a slice? How much of a slice can you buy for $0.25?" and set the students off to talk to each other.

These questions help get at the heart of Proportional Reasoning—what is legal to do with these ratios? How can we maintain equivalence? If you scale the money, do you scale the number of slices? If you subtract some money, do you subtract that same number from the slices? If you subtract some money, do you subtract the corresponding amount of slices? What does it mean to scale in tandem, reasoning proportionally?

Tyler and Pope suggest they could get to that $0.25 by dividing the $2.50 by 10.

That means they have to divide the 2 slices also by 10, resulting in $0.2 = \frac{1}{5}$ of a pizza.

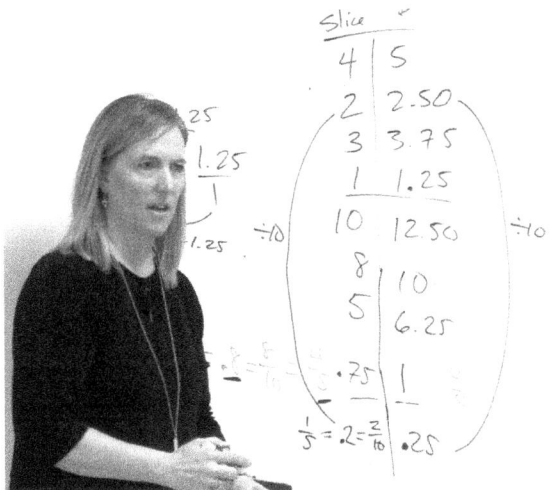

If you divide 2.50 by 10, then you divide the 2 slices by 10, so $0.25 buys $\frac{1}{5}$ of a slice of pizza.

Ah! If you are going to subtract $0.25, you must subtract $\frac{1}{5}$ of a slice, not $\frac{1}{4}$ of a slice. It matters how the slices and money are related.

In this experience, we see Proportional Reasoning being built:

- Two quantities linked and varying together (number of slices of pizza and dollars)

- Scaling in tandem
  - halving the number of slices means you halve the price
  - doubling the price means you double the number of slices
  - scaling the number of slices by any factor means you scale the price by that same factor
  - if you add or subtract a slice, you have to add or subtract the corresponding price for that amount of pizza

- Rational numbers: fractions, ratios, and percents, including:
  - the ratio of 4 slices : $5.00 (a non-unit rate)
  - $1.25 per slice (a unit rate)
  - $\frac{4}{5}, \frac{8}{10}$, 0.8 of a slice (a unit rate)
  - 80% of a slice

To watch these students in action, see the QR code on this page.

Watch students develop proportional reasoning

https://qrs.ly/b4g3rh7

What is Proportional Reasoning? Proportional Reasoning is *"reasoning up and down* in situations in which there exists an invariant (constant or unchanging) relationship between two quantities that are linked and varying together" (Lamon, 2020). This reasoning up and down, scaling, is multiplicative in nature.

FREQUENTLY ASKED QUESTIONS

Q: If I'm telling my students to find equivalent fractions by multiplying the numerator and denominator by the same number, is that scaling in tandem?

A: If you're telling students to do steps, that is not what we mean by the mental action of scaling in tandem. Scaling in tandem is realizing that what you do to the numerator affects the denominator in the same way simultaneously. It's about what is happening in students' minds as they think about scaling. For example, I once asked my son, "What do equivalent fractions mean to you?"

He answered, "If you cut the slices of pizza in half, you have to have twice as many."

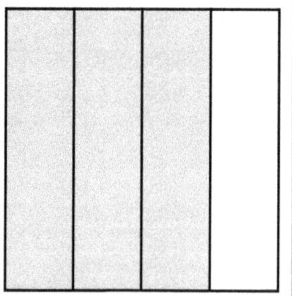
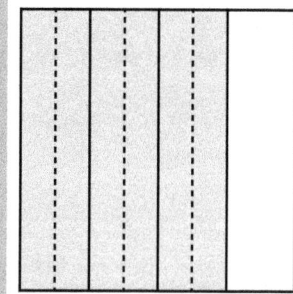

Have 3 slices of pizza out of 4.   Cut each slice in half. Now
Each slice is ¼ of the pizza.   6 slices, each ⅛ of the pizza.

$$\frac{1}{4} = \frac{2}{8}$$

$$\frac{3}{4} = \frac{6}{8}$$

If you have $\frac{1}{4}$ and cut it in half, you now have twice as many pieces. You have to have 2 of those $\frac{1}{8}$-slices to have the same amount, so $\frac{1}{4} = \frac{2}{8}$. If you have $\frac{3}{4}$ of the pizza and cut each in half (2 equal pieces), you have to have $\frac{6}{8}$ to have the same amount. This is scaling in tandem. Fraction equivalence is figure-out-able.

Remember that Lamon estimates that over 90% of adults do not reason proportionally, presenting "compelling evidence that this reasoning entails more than developmental processes and instruction must play an active role in its emergence" (Lamon, 2020, p. 3). This suggests that most people do not develop Proportional Reasoning on their own and would benefit from purposeful experiences to develop this reasoning.

One of the reasons why so many adults do not reason proportionally is they have been trapped in algorithms in the previous reasoning domains (see Figure 5.2).

FIGURE 5.2 ● Proportional Reasoning and the Preceding Domains

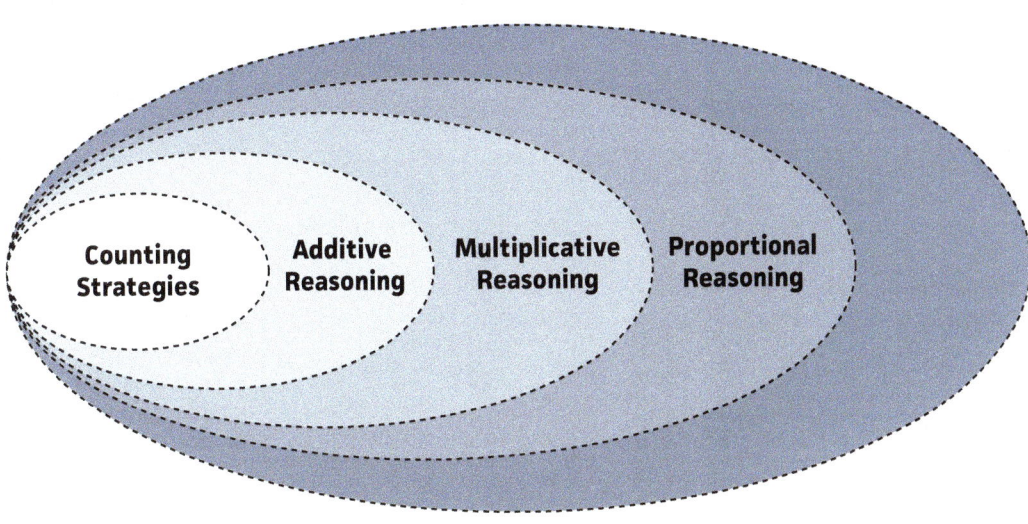

Source: Adapted from Math Is Figure-Out-Able at https://www.mathisfigureoutable.com/ with CC Attribution-NoDerivatives 4.0 International License.

Each important juncture in the development progression asks students to consider more simultaneously, to grapple with more things at the same time. In Proportional Reasoning, students are asked to grapple with everything involved in Multiplicative Reasoning for one number, but also to consider the effect on a different, linked number in all those ways *at the same time*. Not only is a number being scaled (Multiplicative Reasoning), but its partner in the ratio is being scaled in the same way. Table 5.1 shows a comparison of the domains of reasoning.

TABLE 5.1 ● Comparing the Domains of Reasoning

Counting Strategies		Deal with one at a time, many little jumps of one.
Additive Reasoning		Deal with one group at a time, more than one at a time.
Multiplicative Reasoning		Deal with groups of groups, more than one group at a time.
Proportional Reasoning		Deal with groups of groups *compared* to other groups of groups, scaling in tandem, perceiving the groups of red dots in relation to the groups of blue dots.

For the Multiplicative Reasoning row:

1	27
10	270
9	243
5	135

For the Proportional Reasoning row:

×2 {	2	3	} ×2
×2 {	4	6	} ×2
	8	12	
×5 {	40	60	} ×5

FREQUENTLY ASKED QUESTIONS

Q: How is a ratio table in multiplication and division different than in Proportional Reasoning?

A: The only difference between ratio tables used for multiplication and division and ratio tables used to solve proportions is the beginning rate. In multiplication and division, the table starts with a unit rate. In solving proportions, the table usually starts with a non-unit rate.

Another example of the complexity in Proportional Reasoning is the unitizing and re-unitizing required to think about rational numbers. Consider the fraction $\frac{3}{4}$. There is the numerator 3 that is a unit and the denominator 4 that is a unit, but the fraction $\frac{3}{4}$ is a composite unit that refers simultaneously to the 3 and the 4 and the relationship between them. It's almost like you focus on the 3, and on the 4 and on the ratio 3 out of 4; but you also need to see the three-fourths as a number, a value, a location on the number line. And then you can shift focus and think about $\frac{3}{4}$ of *something*, the fraction acting as an operator and referring to yet another unit, $\frac{3}{4}$ of that new unit.

In fact, there are five intertwined, related, yet distinct ways of thinking about rational numbers. The rational number $\frac{3}{4}$ can be thought of with the part-whole relationship, 3 *out of* 4, and operator meaning $\frac{3}{4}$ *of something*. But $\frac{3}{4}$ can also refer to the location on the number line that is three iterations of $\frac{1}{4}$, that $\frac{3}{4}$ is $\frac{1}{4} + \frac{1}{4} + \frac{1}{4}$ and also $3\left(\frac{1}{4}\right)$. We can also consider 3 cookies divided among 4 friends as $\frac{3}{4}$ of a cookie per friend and in this way, it is a quotient and a ratio (Lamon, 2020).

As you read, it's helpful to remember that unitizing *is "treating a quantity as a single unit" (Hackenberg et al., 2016).*

Students deal with even more increasing complexity in fraction multiplication and division. When a student considers a problem like $\frac{3}{4}$ of 16, there are several units to coordinate.

x x x x x x x x x x x x x x x x	(16) one-units	Consider the 16 objects as 16 units. You then have 16 one-units.
x x|x x|x x|x x x x|x x|x x|x x	4 (four-units)	Create units of units— that is, 4 composite units, each consisting of 4 one-units. You then have 4 four-units.
x x|x x|x x|x x x x|x x|x x|x x	3/4 of 16	Create units of units of units—that is, create 1 three-unit consisting of 3 of the 4 four-units.

Credit: Adapted from Lamon (1994).

"Units coordination refers to the ways students build and work with units" (Steffe, 1992, as quoted in Hackenberg et al., Developing Fractions Knowledge, 2016, p. 22). "The goal is to help students develop through the stages of unit coordination where students can take three levels of units as given, and can thus flexibly switch between three-level structures" (Hackenberg et al., 2016, p. 23).

Fosnot and Dolk in their book Constructing Fractions, Decimals, and Percents acknowledge this complexity when they call the fraction operations of multiplication and division "relations on relations" (Fosnot & Dolk, 2002). Reasoning about fraction operations involves a lot of re-unitizing. In this re-unitizing, the whole keeps switching.

TRY IT

Name and describe fractions like $\frac{2}{3}, \frac{3}{5}$, and $\frac{7}{8}$ without using "over." How many helpful ways can you list? See the tip on this page about $\frac{4}{5}$ and the five intertwined, distinct ways listed on page 155 to think about $\frac{3}{4}$.

In this section, we've *listed* and *described* some of the important components of the complex realm of Proportional Reasoning. Truly *developing* Proportional Reasoning, however, requires grappling with problems across the different forms of rational numbers (fractions, decimals, percents) and building the major models (open double number lines, percent bars, ratio tables) and major strategies.

An open double number line model is like an open number line in that one puts tick marks only where they are needed. A double number line represents equivalent ratios at each tick mark. Unlike the more flexible ratio table, the tick marks go in order just like a number line, left to right, smallest to largest.

A percent bar model is similar to an open double number line, but it is used to represent equivalent ratios with percentages. It is open in that one only puts dividing lines where they are needed. Like a number line, the dividing lines represent values in order, left to right, smallest to biggest, but unlike an infinite number line, the bar represents 100% of the total.

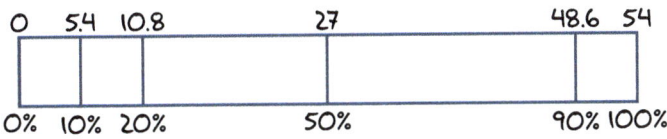

Video of a student solving Proportional Reasoning problems using Additive Reasoning:

https://qrs.ly/5kg3rh9

A ratio table represents equivalent ratios. Unlike an open double number line and percent bar model, ratio table entries do not have to go in order of magnitude. This makes ratio tables very flexible tools.

2	4	40	12	1.2	41.2
3	6	60	18	1.8	61.8

A student demonstrates reasoning about fractions when solving $\frac{8}{7} \times \frac{3}{12}$ as they decide to think about that problem with the factors reversed, as $\frac{3}{12} \times \frac{8}{7}$. Since $\frac{3}{12}$ is equivalent to $\frac{1}{4}$, the problem is like thinking about $\frac{1}{4}$ of $\frac{8}{7}$. The student thinks about $\frac{1}{4}$ of 8 *things* as 2 *things* and $\frac{8}{7}$ as 8 one-sevenths, therefore $\frac{1}{4}$ of 8 one-sevenths is 2 one-sevenths. And 2 one-sevenths is $2 \times \frac{1}{7} = \frac{2}{7}$.

Video of a student solving Proportional Reasoning problems using Proportional Reasoning:

https://qrs.ly/rig3rhb

Reasoning about fractions is also shown in these solutions to problems, like: $\frac{1}{10} + \frac{3}{4}$, $\frac{1}{3} - \frac{1}{4}$, $\frac{1}{4} \times \frac{6}{5}$, and $2\frac{1}{8} \div \frac{1}{8}$.

$\frac{1}{10} + \frac{3}{4} = \frac{2 \text{ nickels}}{20 \text{ nickels}} + \frac{15 \text{ nickels}}{20 \text{ nickels}} = \frac{17 \text{ nickels}}{20 \text{ nickels}} = \frac{17}{20}$ $= \frac{1 \text{ dime}}{10 \text{ dimes}} + \frac{7.5 \text{ dimes}}{10 \text{ dimes}} = \frac{8.5 \text{ dimes}}{10 \text{ dimes}} = \frac{8.5}{10} = \frac{17}{20}$ $= \$0.10 + \$0.75 = \$0.85 \ = \frac{17 \text{ nickels}}{20 \text{ nickels}} = \frac{17}{20}$	Using money to think about common coins/ decimals/ denominators to add fractions and equivalent fractions

(Continued)

(Continued)

$\frac{1}{3} - \frac{1}{4} = \frac{20 \text{ minutes}}{60 \text{ minutes}} - \frac{15 \text{ minutes}}{60 \text{ minutes}} = \frac{5 \text{ minutes}}{60 \text{ minutes}}$ $= \frac{1 \,(5\text{-minute chunks})}{12 \,(5\text{-minute chunks})} = \frac{1}{12}$	Using a clock to think about common chunks of times to find equivalent fractions and to subtract fractions
$\frac{1}{4} \times \frac{6}{5}$ Think $\frac{1}{4}$ is $\frac{1}{2}$ of $\frac{1}{2}$ So, $\frac{1}{2} \times \frac{6}{5} = \frac{3}{5}$ And ½ of 3 is 1.5 so $\frac{1}{2} \times \frac{3}{5} = \frac{1.5}{5} = \frac{3}{10}$	Multiplying fractions using the operator meaning of fractions and equivalent fractions
$2\frac{1}{8} \div \frac{1}{8}$ How many $\frac{1}{8}$s in 1? 8 So, $2\frac{1}{8} \div \frac{1}{8} = 17$ How many $\frac{1}{8}$s in 2? 16 One more $\frac{1}{8}$ so 16 + 1 = 17	Dividing fractions using a quotative approach and equivalent fractions

The examples in the previous section show students reasoning about fractions using relationships between the numerators and denominators of the fractions and also relationships from fraction to fraction.

A student demonstrates Proportional Reasoning when solving the proportion $\frac{15}{2.5} = \frac{12}{x}$ for x as they look to see how they could scale 15 to 12 and scale the 2.5 along at the same time. Multiplying the $\frac{15}{2.5}$ by 4 gives $\frac{60}{10}$, which is convenient because both numerator and denominator are divisible by 5, which yields $\frac{12}{2}$, so $x = 2$.

$$\frac{15}{2.5} = \frac{12}{x} \qquad\qquad \overset{\times 4}{\frac{15}{2.5}} = \underset{\times 4}{\frac{60}{10}} = \overset{\div 5}{\underset{\div 5}{\frac{12}{2}}} \quad \text{so } x = 2$$

Proportional Reasoning is also shown in these solutions to problems like:

$$\frac{14}{8} = \frac{x}{12} \qquad \frac{5}{7} = \frac{6}{x} \qquad \frac{6}{15} = \frac{x}{1}$$

$\frac{14}{8} = \frac{x}{12}$	$\overset{\div 2}{\curvearrowright}\quad\overset{\times 3}{\curvearrowright}$ $\frac{14}{8} = \frac{7}{4} = \frac{21}{12}$ $\underset{\div 2}{\curvearrowright}\quad\underset{\times 3}{\curvearrowright}$ so $x=21$	Scale down to scale up, using the *between* relationship
$\frac{5}{7} = \frac{6}{x}$	$\overset{\times 6}{\curvearrowright}\quad\overset{\div 10}{\curvearrowright}\quad\overset{\times 2}{\curvearrowright}$ $\frac{5}{7} = \frac{30}{42} = \frac{3}{4.2} = \frac{6}{8.4}$ $\underset{\times 6}{\curvearrowright}\quad\underset{\div 10}{\curvearrowright}\quad\underset{\times 2}{\curvearrowright}$ so $x=8.4$	Scale up to scale down, using the *between* relationship
$\frac{6}{15} = \frac{x}{1}$	$\overset{\div 3}{\curvearrowright}\quad\overset{\times 2}{\curvearrowright}\quad\overset{\div 10}{\curvearrowright}$ $\frac{6}{15} = \frac{2}{5} = \frac{4}{10} = \frac{0.4}{1}$ $\underset{\div 3}{\curvearrowright}\quad\underset{\times 2}{\curvearrowright}\quad\underset{\div 10}{\curvearrowright}$ so $x=0.4$	Scale to 10 to get to the unit rate, using the *between* relationship
$\frac{6.25}{12.5} = \frac{3.3}{x}$	$\times 2\left(\frac{6.25}{12.5} = \frac{3.3}{6.6}\right)\times 2$ so $x=6.6$	Use the *within* relationship to scale in tandem

FREQUENTLY ASKED QUESTIONS

Q: Where can I learn more about Proportional Reasoning?

A: To learn more about Proportional Reasoning and to see it in action, check out the module on Proportional Reasoning in my online workshop, *Developing Mathematical Reasoning*, at https://www.mathisfigureoutable.com/dmr/workshop, and check out my upcoming books *Developing Mathematical Reasoning 6–8* (available 2026) and *Developing Mathematical Reasoning 9–12* (available 2027).

DMR Online Workshop

https://qrs.ly/ 6rg3rek

Each of the traditional algorithms for fraction operations falls into the traps of algorithms. In this book, the conversation is limited to the fraction multiplication algorithm and its limitations for building reasoners. The algorithms for the other operations have similar traps.

TIP

Each of these strategies can be employed *within* a ratio, scaling in tandem between the numerators and denominators. Each can also be employed *between ratios*, scaling in tandem from one ratio to the other.

THE TRAP OF A TRADITIONAL FRACTION MULTIPLICATION ALGORITHM

In the United States, textbooks have traditionally chosen a particular fraction multiplication algorithm. The steps of that procedure can be summarized as: Multiply the numerators. Multiply the denominators. Simplify/reduce the result.

This algorithm seems straightforward enough yet inherently has sneaky traps for learners who should be developing Proportional Reasoning with rational numbers.

Let's walk through the example problem $\frac{3}{4} \times \frac{2}{5}$ and consider how a student could get the correct answer while being trapped by the algorithm to never reason about $\frac{3}{4}$ or $\frac{2}{5}$ or their product at all.

To begin this algorithm, the student writes the fractions next to each other and treats the fractions as digits *over* digits.

STEP 1: FOR $\frac{3}{4} \times \frac{2}{5}$, WRITE THE FRACTIONS AND REMEMBER WHICH RULE TO USE WITH MULTIPLICATION

STEP 1 TRAP: ROTE-MEMORY TRAP

If we treat fraction operations as rules that are items to choose from in the menu of fraction algorithms, students can get trapped in an all-you-can-eat buffet of similar-sounding and -looking procedures. A chicken sandwich, chicken salad sandwich, chicken fried steak sandwich—they're all the same, right? Does it really matter which one you eat? How do you remember which one to eat when?

The student stuck in this trap thinks: The operation is multiplication, so choose from algorithms that *sound* like multiplication.

- Multiply across
- Multiply top and bottom by the same number
- Cross multiply
- Cross multiply and divide
- Invert and multiply (skip flip multiply)
- Multiply by the reciprocal
- Multiply then add it's not that bad

- Multiply top and bottom of each fraction by the other's denominator
- Multiply by a fancy (convenient) 1 (multiply or divide by a fraction equivalent to one)
- Top divided by bottom, multiply by 100

Note how none of these fraction algorithms look anything like the "multiplication" students have been doing if they have been constrained to a multiplication algorithm that relies on vertical columns of digits. It is no wonder there is so much confusion when students begin multiplying fractions.

When I shot out a query to Twitter and Facebook for fraction algorithms that have the word *multiply* in them, I received many descriptions of how teachers knew which one went with which algorithm. For example, they would supply the operation and then the algorithm:

Veronica Lucena
Mixed numbers to improper fractions. Multiply the whole by the denominator add the numerator.

Fractions of a set: whole number divided by the denominator, multiply by numerator.

Finding equivalent fractions: if you multiply the denominator you multiply the numerator.

Source: Veronica Lucena

They made the distinction of when to use which rule. Some students can make that distinction. Most do not. I did as a student, but because what was valued was choosing and implementing the correct procedure, I was trapped in correct answers without developing Proportional Reasoning with fractions. And I was not alone. I was part of the 90% of adults who are not reasoning proportionally.

STEP 1, TRAP: DIGIT/LOCATION TRAP "OVER"

In the previous chapters about the four operations with integers, one of the major traps is the digit trap, where larger numbers are not considered for their magnitude because the algorithms require the numbers to be split into digits and then operate on the digits.

With fractions we see the same kind of trap, but this time instead of larger number problems being considered only as columns of digits, here fractions are being considered only as digits *over* digits, not the rational number they represent.

Most, if not all, fraction algorithms require students to break up the fractions into digits *over* digits and operate on those digits.

The word *over* is a positional word—it describes location. The words *over* and *under* describe the position of the numerator and denominator in the way we write a fraction, the look of the notation, but these words do not describe mathematical relationships. We may rarely need to use *over* to describe the notation (for example, for a young learner, "Write the 3 over the 4 to represent three-fourths"), but we should never use it to name the fraction. For example, the fraction $\frac{3}{4}$ should be named *three-fourths, three one-fourths, three-quarters,* or even *three times one-fourth,* but never *three over four.* Using the phrase *three over four* to name $\frac{3}{4}$ can trap students into thinking only about how to write or recognize the way the fraction is written but never considering or grappling with the meaning of $\frac{3}{4}$.

The digit/location trap:	What students could be considering and grappling with:
$3 \qquad 2$ $\frac{3}{4} \times \frac{2}{5}$ $4 \qquad 5$ The digits 3, 2, 4, 5. What do I do with them this time?	$\frac{3}{4}$ *of something* or $\frac{2}{5}$ *of something*
	$\frac{3}{4}$ as three $\frac{1}{4}$s ($3 \times \frac{1}{4}$) or $\frac{2}{5}$ as two $\frac{1}{5}$s ($2 \times \frac{1}{5}$), reasoning about non-unit fractions as multiples of unit fractions
	$\frac{3}{4}$ as the number 0.75 and $\frac{2}{5}$ as the number 0.4
	the area of a rectangle that is $\frac{3}{4}$ of a unit by $\frac{2}{5}$ of a unit
	that $\frac{3}{4}$ of something will always be smaller than that thing (even though up to this point in the students' education, multiplication by whole numbers always meant that the product was larger than the factors)
	$\frac{3}{4}$ as $3 \div 4$ and $\frac{2}{5}$ as $2 \div 5$

STEP 2: MULTIPLY THE NUMERATORS, MULTIPLY THE DENOMINATORS

If students managed to choose or guess the correct algorithm for fraction multiplication from the myriad of choices, now they multiply across.

$$\frac{3}{4} \times \frac{2}{5} = \frac{3 \times 2}{4 \times 5} = \frac{6}{20}$$

STEP 2, TRAP: SINGLE-DIGIT MULTIPLICATION TRAPS

When students do the multiplication of the numerators and denominators, they are at risk of falling into the trap of rote-memory tricks, of using Additive Reasoning (skip-counting), or of using at best single-digit multiplication relationships. So, even if they are using the best of those, single-digit multiplication relationships, students are not grappling with and developing the proportional relationships involved in the problem.

STEP 3: SIMPLIFY/REDUCE THE RESULT IF NECESSARY

Once students have the product, they divide by common factors to find an equivalent fraction. This has been called "simplifying" or "reducing" the fraction and comes with traps.

$$\frac{6}{20} = \frac{3}{10} \quad \text{OR} \quad \frac{6}{20} = \frac{3 \times 2}{10 \times 2} = \frac{3}{10} \times \frac{2}{2} = \frac{3}{10} \times 1 = \frac{3}{10} \quad \text{OR} \quad \frac{6}{20} = \frac{3}{10}$$

STEP 3, TRAP: THE TRAP OF "SIMPLIFY OR YOUR ANSWER IS WRONG"

Because the algorithm includes the step "simplify," we have culturally gotten into the mindset that an equivalent fraction that is not simplified is incorrect and marked as wrong on a student's work. This can trap students into thinking that $\frac{6}{20}$ is not a correct answer to the multiplication question.

If you are finding the number of square units that need to be planted when planting $\frac{3}{4}$ of $\frac{2}{5}$ of your garden, then 6 out of the 20 square units is a fine way to answer. It is also correct that you are planting $\frac{3}{10}$ of the garden. Context matters. The unit, square units or garden, matters.

But even out of context, the solution $\frac{6}{20}$ is *not* incorrect; it's just not in simplified form.

When required to always simplify the solution, teachers and students can also fall into the *cancel* trap.

STEP 3, TRAP: THE CANCEL TRAP

When teachers instruct students to simplify the fraction answer, they can often use the term *cancel*.

Cancel is not a mathematical operation. It is a colloquial term that can mean too many different things with different outcomes.

Wolfram Alpha, an online mathematical dictionary, gives six definitions for the word *cancel*, and not one of them describes a math thing to do (wolframalpha.com, 2024).

The closest definition listed for *cancel*, "to remove or make invisible," is highly problematic in this fraction multiplication/simplify context. Removing connotes subtraction, and making invisible is what students in this trap wish would happen to their homework. Common factors are not being subtracted or made invisible. One helpful way to look at what is really happening is thinking of $\frac{6}{20}$ as $6 \times \frac{1}{20}$ and envisioning bigger, fewer pieces (for an equivalent total). Put the small 6 pieces together to make 3 larger pieces, out of the twice as big pieces as twentieths, tenths.

Cancel?

$$\frac{\overset{3}{\cancel{6}}}{\underset{10}{\cancel{20}}} = \frac{3}{10}$$

What is really happening:

$$\frac{6}{20} = 6 \cdot \frac{1}{20} = 6 \cdot \frac{1}{2} \cdot \frac{1}{10} = 3 \cdot \frac{1}{10} = \frac{3}{10}$$

And *canceling* goes only one direction with equivalent fractions. One cannot find an equivalent fraction with more, smaller pieces by canceling because canceling only produces fewer, larger pieces.

Another form of the algorithm would have students *cancel* as one of the first steps.

$$\frac{3}{\underset{2}{\cancel{4}}} \times \frac{\overset{1}{\cancel{2}}}{5} = \frac{3 \times 1}{2 \times 5} = \frac{3}{10}$$

But what is really going on here? Using the associative and commutative properties, this is factoring the factors, and resulting in $2 \times \frac{1}{4} = \frac{1}{2}$ or the same factor, 2, in the numerator and denominator so they divide to 1.

$$\frac{3}{4} \times \frac{2}{5} = 3 \times \frac{1}{4} \times 2 \times \frac{1}{5} = 3 \times \left(2 \times \frac{1}{4}\right) \times \frac{1}{5} = 3 \times \frac{1}{2} \times \frac{1}{5} = \frac{3}{10}$$

$$\frac{3}{4} \times \frac{2}{5} = \frac{3 \times 2}{4 \times 5} = \frac{3 \times 2}{5 \times (2 \times 2)} = \frac{3}{5 \times 2} \times \frac{2}{2} = \frac{3}{10} \times 1 = \frac{3}{10}$$

This is predicated on generalizing multiplication of fractions. And students should be encouraged to explore what's happening, and why. But to have it as a step and call it *canceling* keeps it in the realm of "things to do," not things to understand and use when reasoning.

When students are caught in the canceling trap, they are not grappling with or thinking about equivalency.

This trap also finds students canceling numbers, no matter the operation involved.

For example, to find $\frac{5}{3} + \frac{27}{2}$, students might recognize the common factor of 3, and they "cancel" those 3s.

$$\frac{5}{\cancel{3}_1} + \frac{\cancel{27}^9}{2} \overset{?}{=} 5 + \frac{9}{2} = 9\frac{1}{2}$$

The result is $5 + \frac{9}{2} = 5 + 4\frac{1}{2} = 9\frac{1}{2}$. But $\frac{5}{3} + 13\frac{1}{2} = 2\frac{2}{3} + 13\frac{1}{2}$, which is way over $9\frac{1}{2}$. You cannot cancel common factors across fractions when adding.

Similar examples:

For this: $\frac{18}{28}$

"cancel" 8s $\frac{18}{28} \overset{?}{=} \frac{1}{2}$

but $\frac{18}{28} \neq \frac{1}{2}$

For this: $\frac{7}{2} - \frac{3}{42}$

"cancel" 7s $\frac{\cancel{7}^1}{2} - \frac{3}{\cancel{42}_6} \overset{?}{=} \frac{1}{2} - \frac{3}{6} = 0$

but $\frac{7}{2} - \frac{3}{42} = 3\frac{1}{2} - \frac{3}{42} \neq 0$

For this: $\frac{2}{3} \div \frac{1}{6}$

"cancel" 2s $\frac{\cancel{2}^1}{3} \div \frac{1}{\cancel{6}_3} \overset{?}{=} \frac{1}{3} \div \frac{1}{3}$

but $\frac{4}{6} \div \frac{1}{6} \neq 1$

STEP 3, TRAP: THE REDUCE TRAP

As bad as the cancel trap is, the reduce trap might be even worse because it works directly against rational number sense.

Remember that the step is to simplify/reduce the fraction that resulted from multiplying.

Proponents of using the term *reduce* can be quick to point out that it should be the whole phrase *reduce to lowest terms*. But I will argue that we need better words than *reduce* and *lowest* to mean

TIP

To see how your students are thinking about canceling, give them these problems. If they cancel, no matter the operation, ask them what "cancel" means to them and ask them when it's allowed.

"find an equivalent fraction that has no common factors in the numerator and denominator."

Reduce in almost any other context means to make smaller, to shrink, to change the magnitude. Used in this context, *reduce* to *lowest* terms connotes to many students to change the value of the fraction to make it smaller. But $\frac{2}{3}$ is not smaller than $\frac{4}{6}$; they are equivalent.

Let's cancel the term *reduce!*

FREQUENTLY ASKED QUESTIONS

Q: Okay, so maybe the term *reduce* isn't great because it sends the message that you're getting a smaller fraction, but is the term *simplify* really all that better? Is a fraction really more simple than its equivalent?

A: The term *reduce* is terrible, but the term *simplify* is not great, either. We really need a better shorthand term to refer to the process of finding an equivalent fraction that has no common factors in the numerator and denominator. Or maybe we should just call it what it is: *Find an equivalent fraction that has no common factors in the numerator and denominator.* In fact, maybe we should say what we mean in other instances in which we use *simplify*, like when we mean *collect like terms,* or *distribute,* or *distribute and collect like terms,* etc.

Q: Why are we requiring students to simplify the answers to fraction problems? Is that necessary, or even helpful?

A: This is a good question. Why do we put so much emphasis on students reporting the most simplified form of a fraction? It's possible that tradition has set in. It could also have to do with student work being easier to grade. We all want students to be able to find equivalent fractions, and this is one place in fraction problems where we can test that. But at what cost? As we shift the focus from answer-getting to reasoning, constructing viable arguments, and critiquing the reasoning of others, the insistence to always simplify an answer will decrease.

Q: But, Pam, students *get* fraction multiplication. Almost all my students can *do* it correctly. We all leave at the end of the class period content and happy, each student merrily multiplying across, slashing and burning, bam. All good. What's the problem?

A: Yes, that class period and maybe even the set of periods while students are multiplying fractions seem to go well. But then the bubble bursts when we add a new algorithm or—even worse—mix the problem types, or—the kiss of death—bring in word problems. Students then mix and match, flip and multiply, find a common denominator, and then cross cancel. It's a minestrone soup where you never know what will be in your spoon.

Q: My students do the box method. They draw a box, cut 1 side into 4 chunks and shade in 3 of them, then cut the other side into 5 chunks and shade in 2 of them, then count the double-shaded squares. What about this method?

A: This is simply a different set of steps that students are mimicking. Notice that students can be counting by ones at each step: cut 1 side into 4 and shade 3, repeat with the numbers on the other side. Then count each double-shaded square, one by one. Students get answers. Students are counting. Is there a place for students to explore the area of a rectangle that measures a fraction by a fraction? Yes. Is it *the* thing to do to get answers? No, because students get trapped.

This trap is particularly insidious for the multiplication and division of fraction algorithms because done in isolation it's just so easy. Why make it harder? Well, because the goal of math class is to develop reasoning, not to get easy answers.

TRY IT

What could it look like to reason about finding $\frac{3}{4} \times \frac{2}{5}$?

- Think about $\frac{1}{4}$ of 2. How could that help you find $\frac{3}{4} \times \frac{2}{5}$?
- Think about $\frac{1}{2}$ of $\frac{2}{5}$. How could that help you find $\frac{3}{4} \times \frac{2}{5}$?
- Think about $\frac{2}{5}$ as a decimal. How could you use that to help you find $\frac{3}{4} \times \frac{2}{5}$?
- Think about $\frac{3}{4}$ as a percentage. How could that help you find $\frac{3}{4} \times \frac{2}{5}$?

How could you use $\frac{1}{4}$ of 2 to help you find $\frac{3}{4} \times \frac{2}{5}$?

If $\frac{1}{4}$ of 2 is $\frac{1}{2}$, then $\frac{1}{4}$ of $\frac{2}{5}$ is $\frac{\frac{1}{2}}{5} = \frac{1}{10}$. And if $\frac{1}{4} \times \frac{2}{5}$ is $\frac{1}{10}$, then scale that by 3 to get $\frac{3}{4} \times \frac{2}{5} = \frac{3}{10}$.

(Continued)

(Continued)

How could $\frac{1}{2}$ of $\frac{2}{5}$ help you find $\frac{3}{4} \times \frac{2}{5}$?

If $\frac{1}{2}$ of $\frac{2}{5}$ is $\frac{1}{5}$, then $\frac{1}{4}$ of $\frac{2}{5}$ is half again, $\frac{0.5}{5} = \frac{1}{10}$. Then scale that by 3 to get $\frac{3}{4} \times \frac{2}{5} = \frac{3}{10}$.

How could thinking about $\frac{2}{5}$ as a decimal help you find $\frac{3}{4} \times \frac{2}{5}$?

Since $\frac{1}{5}$ of a dollar is \$0.20, then $\frac{2}{5}$ of a dollar is twice that, so \$0.40. Now think about $\frac{1}{4}$ of 0.40 is 0.10, so scale that times 3 and $\frac{3}{4} \times \frac{2}{5} = 0.3 = \frac{3}{10}$.

How could thinking about $\frac{3}{4}$ as a percentage help you find $\frac{3}{4} \times \frac{2}{5}$?

Since $\frac{3}{4}$ is 75% and $\frac{2}{5} = 0.4$, think about 75% of 0.4. Since 50% of 0.4 is 0.2 and 25% of 0.4 is 0.1, 75% of 0.4 is $0.30 = \frac{3}{10}$.

FREQUENTLY ASKED QUESTIONS

Q: What are major, important strategies for fraction operations?

A: Some of the important strategies involved in fraction multiplication are shown on pages 157–158, where we show the ways to solve $\frac{1}{10} + \frac{3}{4}$, $\frac{1}{3} - \frac{1}{4}$, $\frac{1}{4} \times \frac{6}{5}$, and $2\frac{1}{8} \div \frac{1}{8}$. These include reasoning about equivalence by scaling in tandem, using the operator meaning of fractions, using the notion that non-unit fractions can be thought of as groups of the unit fraction, $\frac{a}{b} = a \times \frac{1}{b}$, and using both quotative and partitive meanings of division.

THE TRAP OF A TRADITIONAL PROPORTION-SOLVING ALGORITHM

In the United States, textbooks have traditionally chosen a particular proportion-solving algorithm: cross multiply and divide. The steps of that procedure can be summarized as multiply one of the *cross products* and divide that product by the other given term.

This algorithm, deceiving in its simplicity, sets a few enticing traps for learners who should be developing Proportional Reasoning.

Let's walk through the example problem $\frac{4.5}{18} = \frac{x}{27}$, considering how students could compute correct answers while trapped by the algorithm into never reasoning about the ratio of 4.5 to 18 or the fraction $\frac{4.5}{18}$ or the scaling between 18 to 27 or equivalence at all.

STEP 1

For $\frac{4.5}{18} = \frac{x}{27}$, recognize that this problem involves 2 *fraction-looking things* with an equal sign in-between them. This means to choose the appropriate algorithm for this kind of problem, cross multiply and divide.

"Cross multiply and divide"

STEP 1, TRAP 1: THE FALSE DEFINITION OF MATH TRAP AND IDENTITY TRAP

This approach of "recognize which kind of fraction/ratio problem this is and choose the correct procedure" feeds right into the notion that math is a set of arbitrary procedures to rote-memorize. Don't think—pull from rote-memory. Don't worry about why it works or how it works. Just choose the correct steps and do them. Students who don't memorize well are likely to choose the wrong procedure. Students who want to understand are often thwarted because the reasoning in the steps is so opaque. Students who memorize well are trapped into believing they are good at math, when in reality they are good at rote-memorizing and mimicking.

We see this trap at work as well-meaning teachers help students rote-memorize this algorithm with nicknames/mnemonics like the bat and ball, butterfly, or criss-cross method.

Bat and ball

Butterfly

Students can become so divorced from meaning that if they accidentally reach for the wrong algorithm—for example, using "multiply straight across"—they will get a fraction answer that is the product of the two fractions instead of finding the missing numerator.

A proportion is a comparison of two equivalent ratios, $\frac{a}{b} = \frac{c}{d}$.

STEP 1, TRAP 2: PIECES TRAP

To *cross multiply and divide* requires the student to treat the proportion as unrelated pieces to do things with. This is similar to the digit traps of the four operation algorithms. Pick apart the proportion into bits and operate on the bits.

Students are thus enticed to focus on *bits and doing* rather than on each ratio as a relationship and the fact that the two ratios are equivalent. You might hear people say, "But it's so much easier to focus on *bits and doing* rather than grapple with and make sense of the ratio relationships." And we get what we've gotten—students who focus on *bits and doing* and never develop Proportional Reasoning.

STEP 2: MULTIPLY

Multiply the numerator of one ratio by the denominator of the other ratio.

$$\frac{4.5}{18} = \frac{x}{27}$$

Multiply 4.5 × 27.

STEP 2: ALL OF THE TRAPS OF THE TRADITIONAL MULTIPLICATION ALGORITHM

If the student does that multiplication using a traditional algorithm *all* of those traps accompany us here, including the turtle laying a magic egg (as shown) and the butt-cheek method. We will not repeat them all here (refer to Chapter 4 for more).

$$\frac{4.5}{18} = \frac{x}{27}$$

Multiply 4.5 × 27.

$$
\begin{array}{r}
\overset{\scriptstyle 1}{}\overset{\scriptstyle 3}{} \\
4.5 \\
\times\ 2\,7 \\
\hline
3\ 1\ 5 \\
9\ 0\ 0 \\
\hline
1\ 2\ 1.5
\end{array}
$$

Turtle

← lays the magic egg

Don't forget to butt-cheek!

STEP 3: DIVIDE

Divide the previously found product by the remaining number.
In this example, $121.5 \div 18$.

$$\frac{4.5}{18} = \frac{x}{27}$$

Multiply 4.5×27.

Divide $121.5 \div 18$

$$\begin{array}{r} \overset{1}{\overset{3}{4.5}} \\ \times\ 2\,7 \\ \hline 3\,1\,5 \\ 9\,0\,0 \\ \hline 1\,2\,1.5 \end{array}$$

Turtle

← lays the magic egg

Don't forget to butt-cheek!

STEP 3: ALL OF THE TRAPS OF THE TRADITIONAL DIVISION ALGORITHM

If the student does that division using a traditional algorithm *all* of those traps accompany us here, including Dirty Monkeys Smelling Bad. Again, refer to Chapter 4 for a clear delineation of the many traps lying in wait here.

$$\frac{4.5}{18} = \frac{x}{27}$$

Multiply 4.5×27.

$$\begin{array}{r} \overset{1}{\overset{3}{4.5}} \\ \times\ 2\,7 \\ \hline 3\,1\,5 \\ 9\,0\,0 \\ \hline 1\,2\,1.5 \end{array}$$

Turtle

← lays the magic egg

Don't forget to butt-cheek!

Divide $121.5 \div 18$.

$$\begin{array}{r} 6.7\,5 \\ 18\overline{)1\,2\,1.5\,0} \\ -1\,0\,8\downarrow \\ \hline 1\,3\,5 \\ -1\,2\,6\downarrow \\ \hline 9\,0 \\ -9\,0 \\ \hline 0 \end{array}$$

Dirty Monkeys Smell Bad

STEP 4: ACKNOWLEDGE THE ANSWER

The answer is the quotient from the division problem, $x = 6.75$.

STEP 4: PIECES TRAP REVISITED WITH THE PREVIOUS (LESS COMPLEX) REASONING

In this algorithm, students are in the split-the-proportion-into-pieces-and-operate-on-the-pieces mode. The operations are multiply and divide. Nowhere are students considering the proportion, the relationships within each ratio and between the ratios. They are not grappling with or developing Proportional Reasoning at all. This represents a lost opportunity to consider the ratios in tandem and look for proportional relationships.

TRY IT

What could it look like to reason proportionally about solving $\frac{4.5}{18} = \frac{x}{27}$?

- Think about $\frac{4.5}{18}$. Is there an equivalent fraction that occurs to you? How could an equivalent fraction help find x?

- What is the relationship between 4.5 and 18? How could you scale from 4.5 to 18 or from 18 to 4.5? How could you use that relationship to reason about scaling to or from 27?

- Think about the relationship between 18 and 27. How could you scale from 18 to 27, maybe scale up to scale down or scale down to scale up? How could finding the scale factor from 18 to 27 help?

How could you use a fraction that is equivalent to $\frac{4.5}{18}$ to help?

Since $\frac{4.5}{18} = \frac{9}{36} = \frac{1}{4}$, then $\frac{4.5}{18} = \frac{1}{4} = \frac{x}{27}$. This means that since $4 \div 4 = 1$, then $27 \div 4 = x$. So, $x = \frac{27}{4}$. We can leave that or find that $\frac{27}{4} = \frac{24}{4} + \frac{3}{4} = 6 + 0.75 = 6.75$. So, $x = 6.75$.

How could scaling within 4.5 and 18 help?

$$\frac{4.5}{18} = \frac{x}{27} \quad \text{so} \quad \frac{18}{4.5} = \frac{27}{x}$$

$$\div 2 \left(\begin{array}{c|c} 18 & 27 \\ 9 & 13.5 \\ 4.5 & 6.75 \end{array} \right) \div 2 \qquad \text{So } x = 6.75$$

How could scaling between 18 and 27 help?

$$\overset{\div 2}{\frown} \quad \overset{\times 3}{\frown}$$
$$\frac{4.5}{18} = \frac{2.25}{9} = \frac{6.75}{27} \qquad \text{So } x = 6.75$$
$$\underset{\div 2}{\smile} \quad \underset{\times 3}{\smile}$$

SOLVING PROPORTIONS IN GENERAL

What about solving difficult proportions? Or solving proportions in general?

Students who are reasoning while scaling in tandem can look at any proportion and note that they need to find the scale factor to get from a to c and then use that scale factor to scale b to x.

Find this scale factor.

$$\frac{a}{b} = \frac{c}{x}$$

Use it to scale b to x.

They can then reason to find that scale factor by dividing $c \div a = \frac{c}{a}$.

Now they scale b by $\frac{c}{a}$. So, $x = \frac{c}{a} \cdot b$.

$$\frac{a}{b} = \frac{c}{x} \qquad \text{So, } x = b \cdot \frac{c}{a}$$

It is noteworthy that when students are reasoning to solve a proportion, they often divide $c \div a$ first, then multiply by b, even though the algorithm says to multiply first and then divide.

Remember, it's not about getting answers. It's about building reasoning.

Conclusion

Early in my work, when my kids were still young, I did most of my reading, experimenting, and work with teachers and students using whole numbers. I had yet to read or try much with fractions.

I was keenly aware of all the "rules" for fractions. At that point, my ability to work with fractions was completely constrained within these algorithmic procedures. I used them fairly frequently as part of the high school classes I taught, yet I had no ability to reason outside the very narrow bounds created by these memorized procedures.

Through some Professional Development I was facilitating for leaders in Texas I met Garland Linkenhoger, who at the time was a district math specialist. A former teacher, she had the unique background of teaching kindergarten in the morning and then crossing the street to teach calculus in the afternoon. Talk about a unique K–12 perspective!

One hot, summer day as we walked into my kids' school to do a joint training with the elementary teachers, Garland was lamenting about a session at a recent national conference. We were discussing the ramifications of the presenter's limited understanding about ratio and fractions, and frankly, I was having a hard time keeping up. My understanding had been so limited by being trapped in the fraction algorithms.

Garland made an outlandish statement that couldn't be true. She said, "You know, Pam, if one truly understands fraction equivalence, you do not need any fraction algorithms."

I stopped walking, turned to her, and said, "What?"

She began explaining what she meant and I politely put my hand up to ask her to pause. "Please just repeat what you said."

"If we actually understand, really make sense of, and get fraction equivalence, we do not need any fraction algorithms. You can reason through any fraction problem."

How? What? I couldn't even. I could not conceive of solving fraction problems without algorithms. I had been trapped in the algorithms.

I am not trapped anymore. You and your students do not have to be either.

Discussion Questions

1. How did you think about fractions and solving proportions as a student? How has your experience influenced the way you have taught?

2. For these problems: $\frac{7}{10} + \frac{1}{4}$, $\frac{2}{3} - \frac{1}{4}$, $\frac{5}{15} \times \frac{9}{12}$, $1\frac{1}{7} \div \frac{1}{7}$, $\frac{3.5}{7} = \frac{x}{24.5}$

   a. How do you reason through them?

   b. Predict how your students are reasoning about these problems.

   c. Ask students an appropriate problem or two. How did they respond?

   d. How might you make your or your students' thinking visible?

3. How are your students reasoning about fraction situations/ problems? Using algorithms, in which they are reasoning additively or multiplicatively? Reasoning proportionally by scaling in tandem? To help you decide, you could use the "How are you thinking about fraction operations right now?" questions on page 143 or the "How are you thinking about solving proportions right now?" questions on page 144.

4. What is a unit fraction and a non-unit fraction, and how are they related?

5. What are all the ways you can think about the rational number $\frac{2}{5}$? Compare your thoughts to the discussion of $\frac{3}{4}$ on page 155. Which are you still working to make sense of and use in your reasoning?

6. How has your Proportional Reasoning developed while reading this chapter?

7. How have your ideas about developing Proportional Reasoning with your students changed while reading this chapter?

TRY IT IN YOUR CLASSROOM

As Close as It Gets

Rational numbers are often very rule-bound for students. They look at anything that resembles a thing-over-a-thing and they start rolling a die for which rule to apply. Without any sense of what is going on, students often feel defeated before they begin. Then, when they do begin, they are not beginning with any sense of magnitudes, how the numbers are related, if a percent or decimal representation might be easier to work with, and the list goes on. Instead, they focus on doing things: flip, find a common denominator, multiply across, cross cancel, etc. An understandable reaction, given the traps of algorithms, but a reaction they must unlearn.

We can help students realize they can kick their "must mimic a rule" dependency by using an instructional routine like As Close as It Gets. This routine helps students focus on using their intuition and what they know in order to reason.

The routine consists of asking students a multiple-choice question with a clever twist—none of the answers is the correct answer. The goal is to choose the answer that is as close as it gets and defend your thinking. Students get to compare their reasoning with other

(Continued)

(Continued)

students in a low-stakes environment. They can ask for clarification and realize that fractions are figure-out-able!

For example, with a question like:

$\frac{8}{9} + \frac{9}{8}$	
a. 0	b. $\frac{1}{2}$
c. 1	d. 2

students could reason:

- $\frac{8}{9}$ is about 1 and $\frac{9}{8}$ is about 1. So, about 1 and about 1 is about 2.

- $\frac{8}{9}$ is a little under 1 and $\frac{9}{8}$ is a little over 1. So, a little under 1 and a little over 1 is about 2.

If students are stuck, teachers can ask:

- What do you know about $\frac{8}{9}$? Why?

- How can you use $\frac{1}{9}$ to talk about $\frac{8}{9}$? And $\frac{9}{9}$ to talk about $\frac{8}{9}$?

Purpose

Encourage reasoning, not mimicking. Because students are looking at the whole fraction—not just the number on top and the number on the bottom, but at the relationship those numbers create—students start to make sense of the size of the fraction. Students can relax into considering how the fractions relate to each other.

Routine

- Tell (remind) students that the correct answer is not listed. Just get as close as they can.

- Show the problem.

- Ask students to defend their answer choice.

- Discuss the reasoning.

Four samples:

$\frac{5}{11}+\frac{4}{7}$	$\frac{21}{22}-\frac{7}{15}$	$\frac{4}{9}\times\frac{12}{13}$	$\frac{11}{5}\div\frac{12}{23}$
a. 0 b. $\frac{1}{2}$	a. 0 b. $\frac{1}{2}$	a. 0 b. $\frac{1}{4}$	a. 0 b. 1
c. 1 d. 2	c. $\frac{3}{4}$ d. 1	c. $\frac{1}{2}$ d. 1	c. 2 d. 4

Important to Consider

If students dive right in, starting to do steps of procedures, remind them the correct answer is not listed. The goal is to just get an answer that is as close as it gets.

Ask questions like:

- Can you picture this fraction?

- Where is the fraction on a number line? What friendly number is this fraction close to? How do you know? How could you use that to help you?

- Can you use what you know about these fractions to reject any of these answers?

Sequencing. Use similar items at the beginning, middle, and end of a unit. This allows you and your students to see and feel growth. Also, begin by choosing problems that are more obvious (the fractions are very close to a landmark like $\frac{1}{2}$, 1, 2) like $\frac{5}{11}$, $\frac{7}{8}$, $\frac{17}{9}$. When students are starting to get confident with those fractions, move to less obvious choices (less typical fractions close to $\frac{1}{2}$ and 1 but also close to other landmarks like $\frac{1}{5}$, $\frac{1}{4}$, $\frac{3}{4}$, $1\frac{1}{2}$) like $\frac{47}{100}$, $\frac{33}{37}$, $\frac{19}{100}$, $\frac{26}{100}$, $\frac{74}{100}$, $\frac{16}{5}$. Use fractions and decimals within fractions, like $\frac{4.6}{9}$, $\frac{\left(2\frac{3}{8}\right)}{5}$. Use items like $-\frac{7}{8}+\frac{8}{7}$ to get students reasoning about operations with integers.

(Continued)

(Continued)

Extension

Design the distractors a–d to be closer to the actual answer so students are nudged to use more detailed reasoning. Ask students to write their own a–d answers based on what they think a good set of distractors would be.

You can use this routine with any operation and size of numbers. Try it with addition, subtraction, multiplication, division, solving proportions, finding slope, etc.

Thank you to Tierney and Russell (2001) for the inspiration in their "Nearest Answer" activity in *Ten-Minute Math*.

CHAPTER 6

Lost in Functional Reasoning

FIGURE 6.1 ● Functional Reasoning is the last domain in the K–12 reasoning progression.

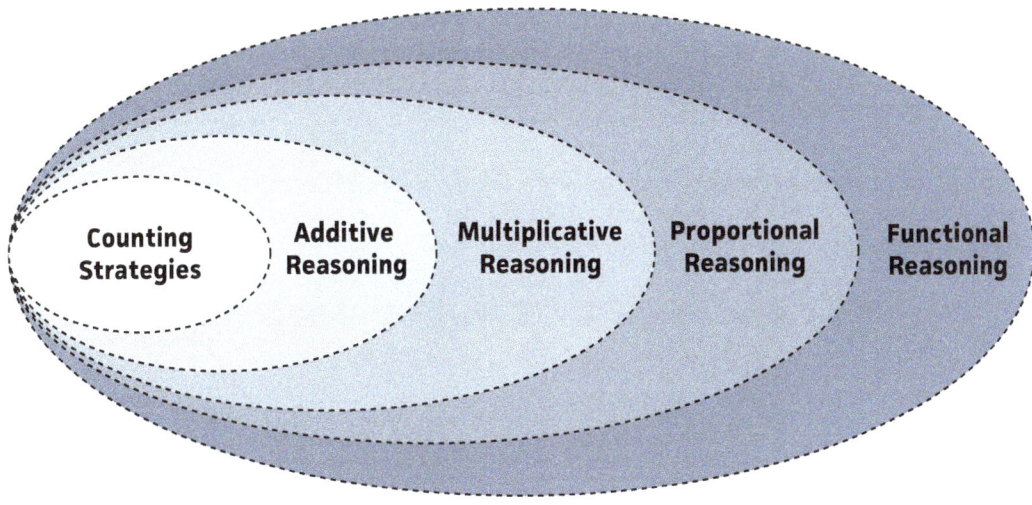

Source: Math Is Figure-Out-Able at https://www.mathisfigureoutable.com/ with CC Attribution-NoDerivatives 4.0 International License.

Graphing calculators and I both started appearing in classrooms at about the same time. I was blessed to have two mentors, Pam Giles (cooperating teacher) and Scott Hendrickson (department chair in my first high school), who were both committed to using technology appropriately as a tool. We were on the cutting edge. For my second year teaching, Scott and I bought new textbooks based on the power of visualization and checked out TI-82 graphing calculators to each of our precalculus students.

The textbook was Precalculus Mathematics: A Graphing Approach, *by Demana and Waits.*

It is early in the year and I have done all of the later problems in the homework set to make sure I am ready for class that day. At the start of class, students all ask about #2. No problem. As it is one of the first of the large problem set, I am confident it will be easy for me.

The problem: Find an appropriate viewing window for the graph of $y = 100\sqrt{x}$.

Before you continue, predict what that function looks like and what an appropriate window in a calculator would be. Grab a handheld or use desmos.com and find a window you like. Then read on.

An appropriate viewing window is a window in which you can see the important features of a graph. Sometimes an appropriate viewing window could be a square window, where the horizontal distance between tick marks is the same as the vertical distance between tick marks.

Reaching for the handy projected graphing calculator, we type it in and see in the default [–10, 10], [–10, 10] window, well, nothing.

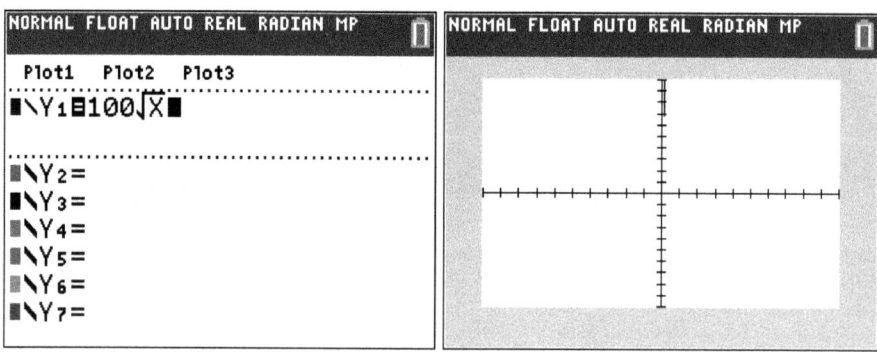

Source: Graphs generated by TI-84CX, Texas Instruments.

I smile confidently and hit the zoom out key, which shows [–40, 40], [–40, 40].

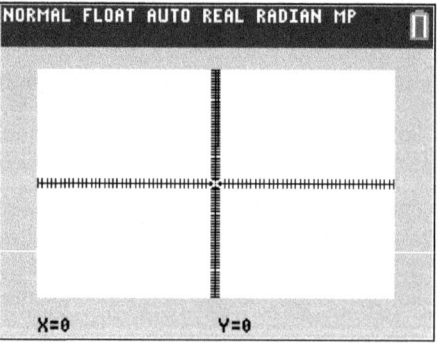

Source: Graphs generated by TI-84CX, Texas Instruments.

Hmmmm....

We get rid of the tick marks by setting the scales to zero.

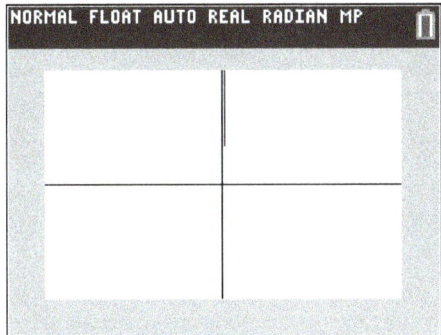

Source: Graphs generated by TI-84CX, Texas Instruments.

I'll admit, I'm not sure what to think or try. In the moment, I don't really know what a square root parent function looks like, nor what a vertical stretch of 100 would do to it. My schooling has included neither parent functions nor transformations.

"What is anyone else thinking?" I ask.

After a minute of everyone trying to find a better window on their own calculators, Carrie says she has hit the *trace* key and found a point.

TIP

When you are unclear, it is a great teacher move to ask, "What do you think?" and let students think *with* you. Play that card early and often when you're clear on what's happening and then it comes in handy when you are not as clear. Students rise to the occasion! Fantastic conversations and learning together can be the result.

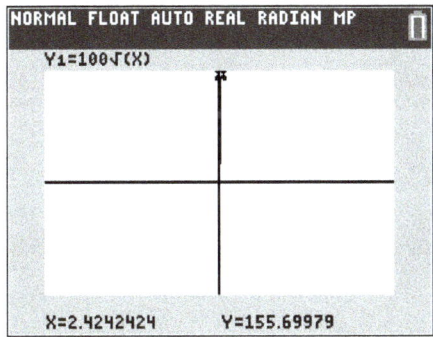

Source: Graphs generated by TI-84CX, Texas Instruments.

That point, approximately (2, 155), helps us think about opening the window to look up.

Which, in turn, helps us to keep adjusting the window.

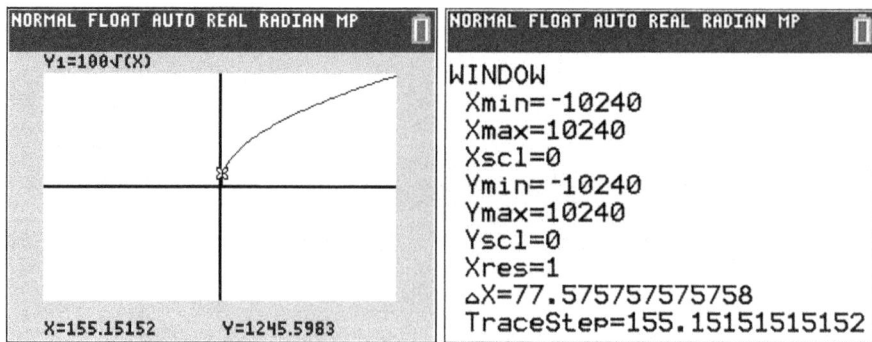

Source: Graphs generated by TI-84CX, Texas Instruments.

Right! That's what a square root function looks like. But look at that window! We have a super conversation about why that window. First, we know that the square root of 0 is 0, so we have the point (0, 0). To find another point on the graph, find the square root of 1, which is 1, but multiply that by 100 for the point (1, 100). That function rises quickly from the origin! And $100\sqrt{4} = 100 \cdot 2 = 200$, for the point (4, 200).

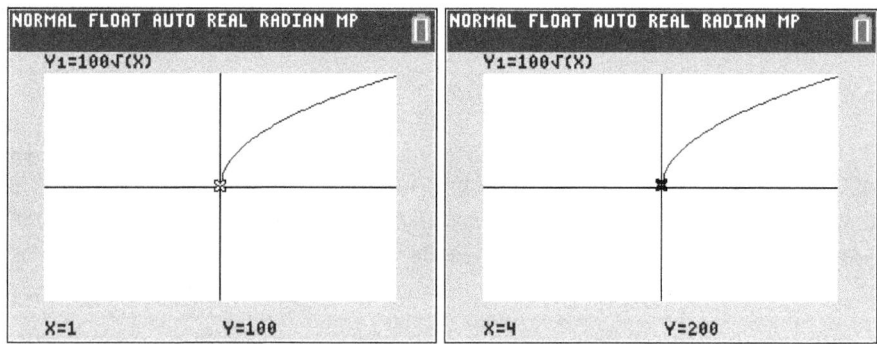

Source: Graphs generated by TI-84CX, Texas Instruments.

And then the function slows down. We look at other points like (1000, 3162) and (10000, 10000). The y-values are rising, but slower than before.

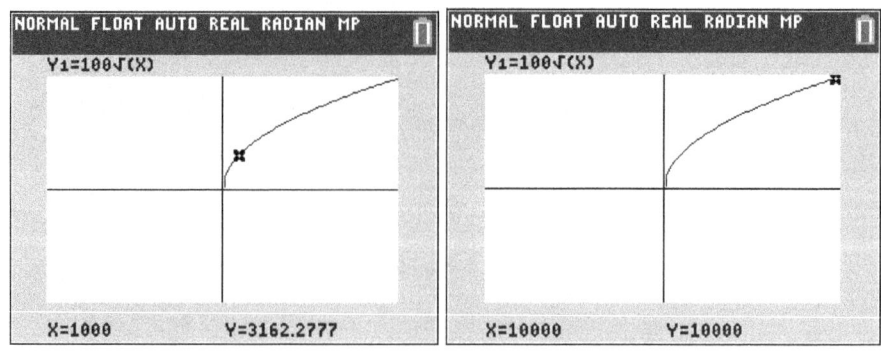

(quick rise)   (slow down)

Source: Graphs generated by TI-84CX, Texas Instruments.

For fun, we compare it to $y = \sqrt{x}$ in the window [−10, 10] by [−10, 10]. Windows matter, but the overall shape of the graph is the same. Interesting!

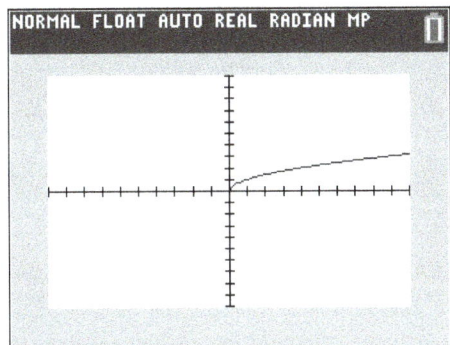

Source: Graphs generated by TI-84CX, Texas Instruments.

We are reasoning about domains and ranges, parent functions, transformations, individual points, and sets of points that could be treated as an object. We are learning to reason functionally.

TIP

Finding appropriate viewing windows using technology can also help build other important concepts. As students are changing the window, the shape of the function may be stretched or compressed, but the overall pattern stays the same. Some rates of change are constant $(y = x, y = |x|)$, and many are not $(y = x^3, y = 2^x)$. Some rates grow slowly $(y = \log x, \ y = \sqrt{x})$, while others grow quickly $(y = e^x, y = \tan x)$. Of course, all of that depends on the domains in which you're looking. Asymptotes "appear" in some windows and not in others—a situation ripe for investigation. See the end of this chapter for more suggestions.

HOW ARE YOU THINKING ABOUT FUNCTIONS AND RELATIONS RIGHT NOW?

Was your instruction about functions and relations primarily algorithms? How do *you* think about functions and relations now?

In this section, focus on your instincts. How does your brain want to attack this scenario? Solve the problem, then read the descriptions underneath, and choose one that best fits your thinking.

How do *you* think about this situation? Your club is selling T-shirts. The setup fee for the booth is $200 and you'll charge $15 a shirt.

- Do you think about finding the cost for one T-shirt and then add a T-shirt at a time to find a few more costs? If you start with $200 and repeatedly add $15, you are using a recursive additive strategy.

- Do you think about some multiples of $15 and add to 200? If you find some groups of shirts and add the $200 to get a few costs, you are using a multiplicative strategy.

- Do you think about the rate of $15 per one shirt so the revenue in general is $r(t) = 15t$ (Proportional Reasoning) and realize you won't start making money until you earned $200, so you think about where $r(t) = 15t = 200$? If you think about the whole scenario and consider some points individually, you are reasoning functionally.

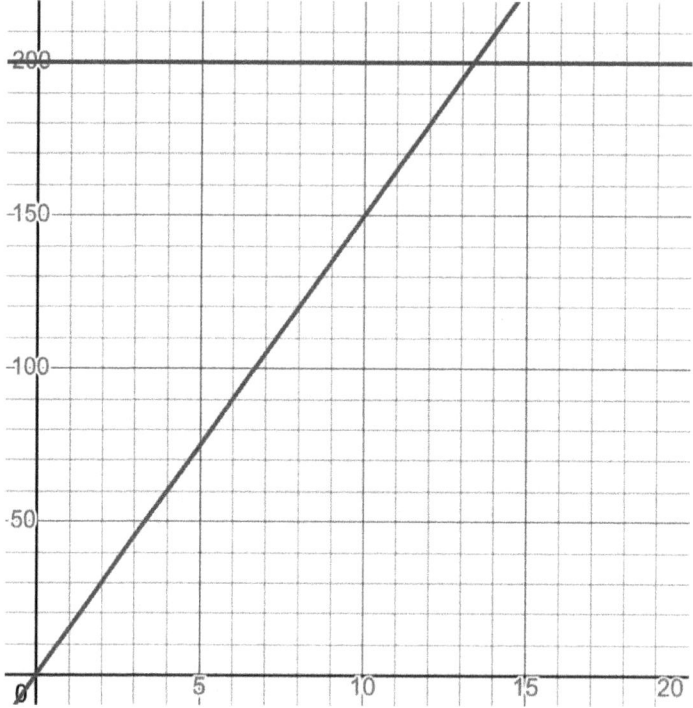

- Do you think about the profit in a general way, that $p(t)$ is to start with a debt of 200, –200, and add $15 per shirt (Proportional Reasoning), $p(t) = -200 + 15t$? Now that you've thought about the whole scenario, you consider a few important points. Since you start in debt, that's the point $(0, -200)$. You can visualize a break-even point where the line crosses the x-axis, at $200 \div 15$, around 13 shirts. If you visualize the entire scenario and consider some points individually, you are reasoning functionally.

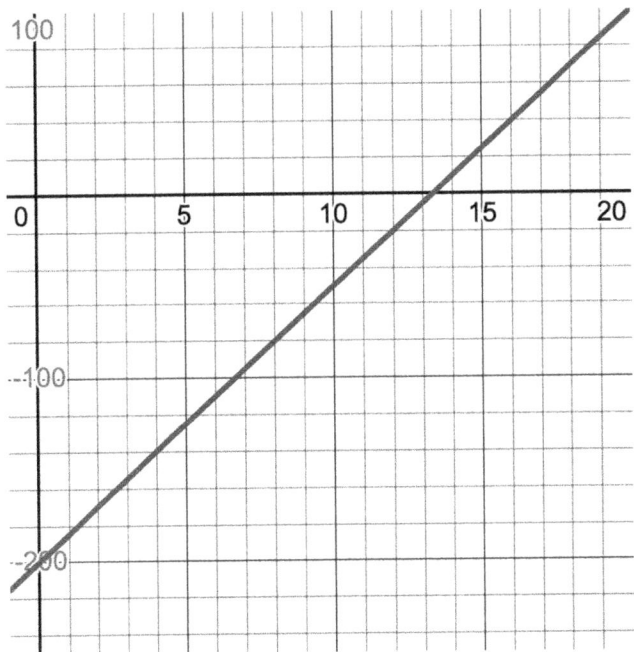

Source: Attributed to Desmos: Creative Commons "Attribution-ShareAlike" license (CC-BY-SA-4.0).

How do you think about sketching a graph of a rational function like $r(x) = \dfrac{3x^3}{x-2}$?

- Do you reach into memory and see a list of things to do? Set numerators and denominators to 0, plot those things (try to remember which is a point and which is an asymptote), look for the coefficients of the largest terms from the numerator and denominator and graph a dotted line with that slope, $y = \dfrac{3}{1}x$? With those points plotted and the line graphed, connect the points, avoid the line. If you follow a series of steps like this in a prescribed order, you might have used no mathematical reasoning—Additive, Multiplicative, or Proportional—in each of those steps.

- Do you see a ratio of two polynomials, considering how the long and short run behavior of those polynomials affects $r(x)$? You compare the long run behavior of the numerator $y = 3x^3$ to the long run behavior of the denominator $y = x - 2$ and realize that $\dfrac{3x^3}{x-2} \approx \dfrac{3x^3}{x} = 3x^2$. Using what you know about $y = 3x^2$, you now have a sense of the long run behavior of $r(x)$. Considering the short run, you think about the short run behavior of each polynomial: the numerator, $3x^3$, has a zero at $(0, 0)$ so $r(x)$ also contains $(0, 0)$ and $r(x)$ is undefined at $x = 2$ and so $r(x)$ has a vertical asymptote at $x = 2$. If you are reasoning, using what you know about the ratio of polynomials, you are using Proportional and Functional Reasoning.

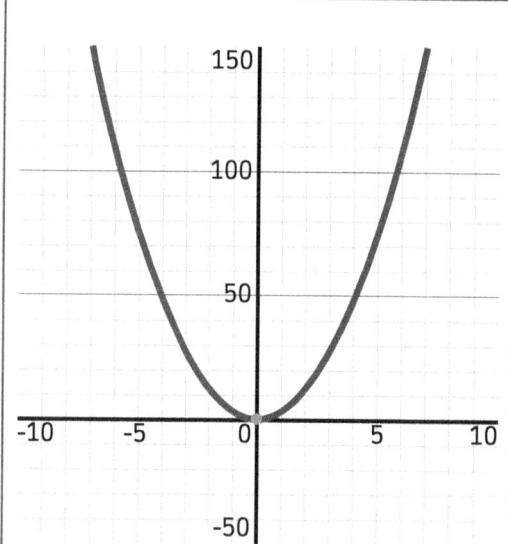

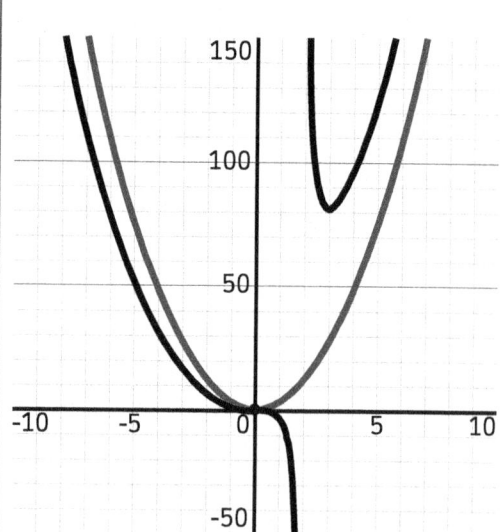

Use the graph of $y = 3x^2$

To inform the end behavior of $f(x) = \dfrac{3x^2}{x-2}$

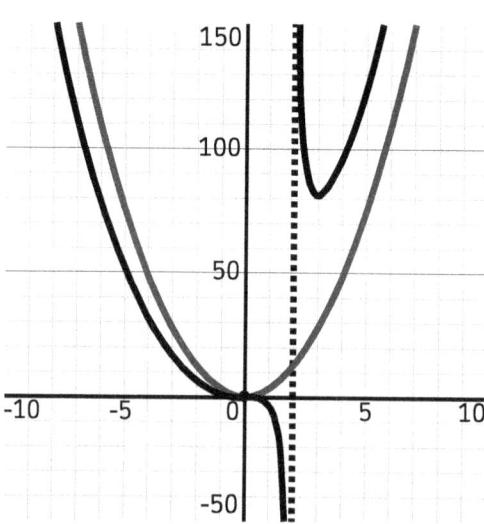

Use the short run behavior of the numerator and denominator to inform the short run behavior $(0, 0)$ and asymptote at $x = 2$.

I first learned to reason functionally about rational functions from the book Functions Modeling Change, *where the authors write: "Mathematics has the extraordinary power to reduce complicated problems to simple rules and procedures. Therein lies the danger in teaching mathematics. It is possible to teach the subject as nothing but the rules and procedures—thereby losing sight of both the mathematics and of its practical value" (Connally et al., 2000).*

FREQUENTLY ASKED QUESTIONS

Q: How can we possibly reason about rational functions if we have students who are still using Counting Strategies for fractions?

A: While this is not trivial, nor easy, we can do some things. This is where a good multiple-access Problem String can help. All throughout your course, facilitate some Problem Strings that build fraction and ratio sense. If you haven't by the time you get to rational functions, do it then. A possible Problem String to help is:

$$1 \div 2$$
$$1 \div 4$$
$$1 \div 10$$
$$1 \div 1000$$
$$1 \div 1$$
$$1 \div \frac{1}{2}$$
$$1 \div \frac{1}{4}$$
$$1 \div \frac{1}{10}$$
$$1 \div \frac{1}{1000}$$

Plot points to represent $(x, \frac{1}{x})$ for each of those questions. Then repeat, but with the opposite of each x value, like $1 \div -2$ for the point $(-2, -\frac{1}{2})$.

Talk about the division, the resulting fraction, the points, the trends. Help students develop both fraction sense and a feel for the rational parent function $y = \frac{1}{x}$.

WHAT IS FUNCTIONAL REASONING?

"What does the graph of $y = x^2$ look like?" I ask a group of high school students taking dual-credit college algebra. I am the guest teacher, filming example Problem Strings. Some students think the graph would be a line, as you can see by the students' body language in this picture.

Sarah argues for a more V shape.

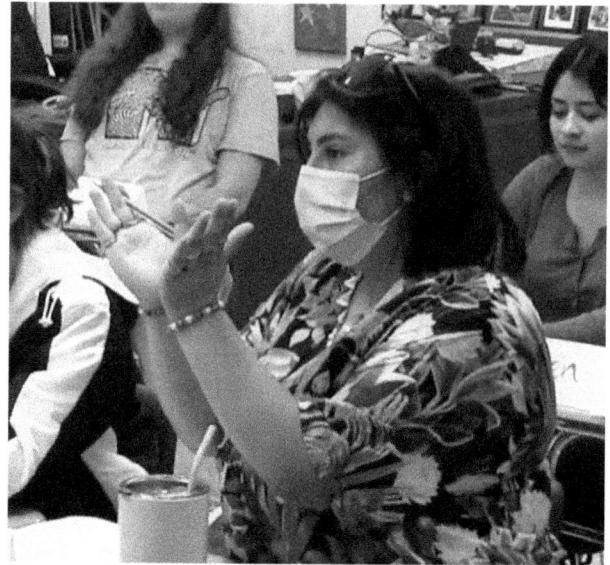

Colby suggests that if we are squaring numbers, the "negatives would be positives" and so the y-values would be positive. With the students' input, we plot some points and connect them. Sure enough, in the second quadrant, the y-values are positive because we're squaring.

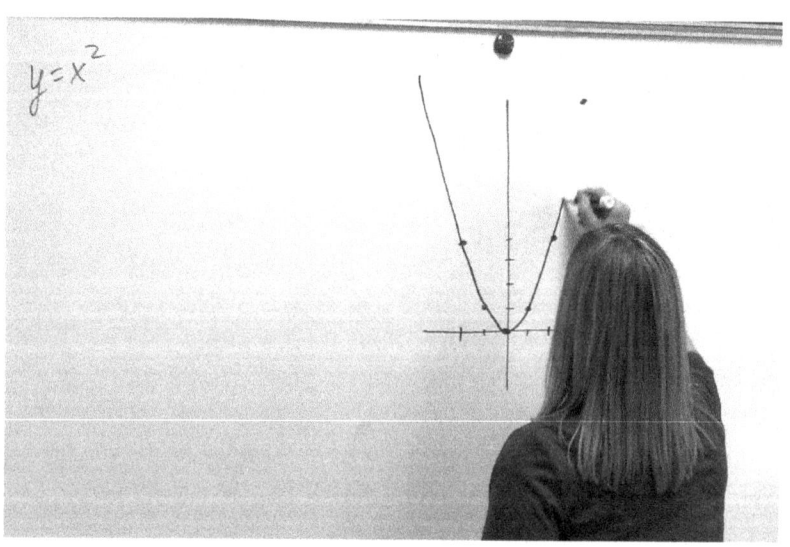

"Next question. What does $y = 3x$ look like?"

A couple of students say, "Wider." After a brief discussion between Brooke and Colby that this function does not have any squaring happening, so instead the negative xs would result in negative ys, we get a graph of the line added to the board. (It might be the worst line I've ever drawn freehand on a board. Sigh.)

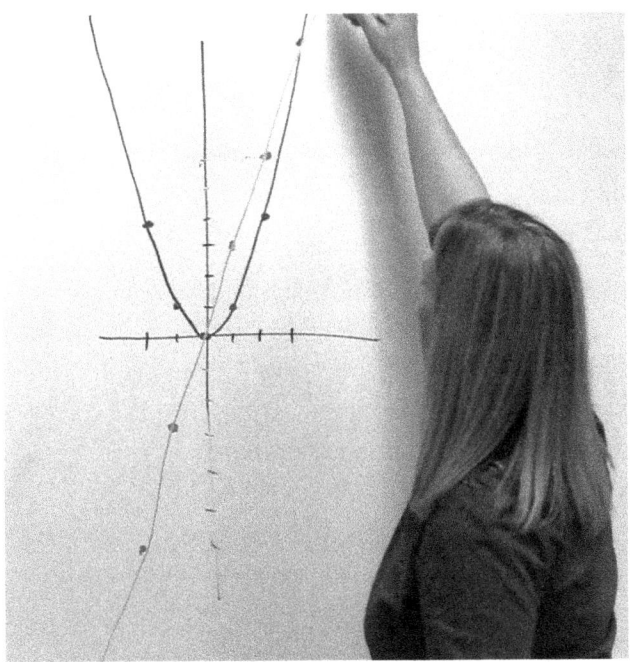

I then ask the next question, $y = x^2 + 3x$. "What would that look like? What does it even mean to add a parabola and a line?"

I ask them to turn and talk to a partner. As I listen in, I hear things about rules, transformations, V-looking graphs, parabolas, lines, and lots of numbers.

Devonte and Caroline discuss the importance of $x = -3$, that it means the combined function would "start" there. This will come back up at the end of the String.

"I see some of you thinking about plugging in some points and then plotting points. I have a question. If we were to really think about this parabola plus that line, could we actually use this graph to do what you were doing? Plug 1 in and find out what it is for the parabola and find out what it is for the line, add 'em together and get the result? Would you all do that with me?"

"If we're at $x = 1$, what is the height of the parabola?	"What is the height of the line?"	"What's the height of the combined function?"
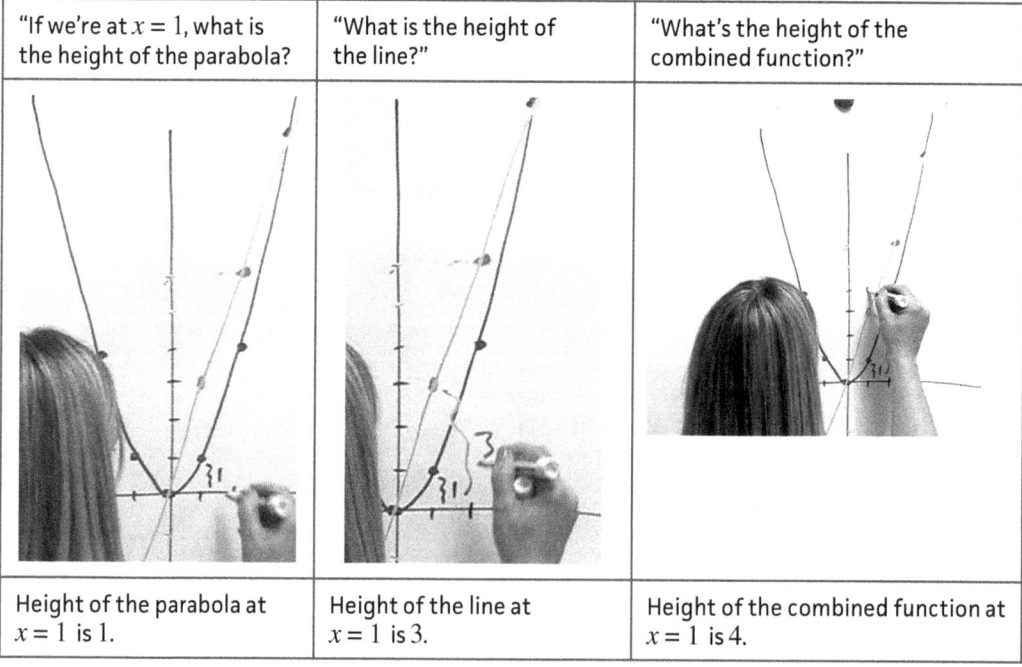		
Height of the parabola at $x = 1$ is 1.	Height of the line at $x = 1$ is 3.	Height of the combined function at $x = 1$ is 4.

"So, if you take the parabola at 1 and add it to the line at 1, you guys are telling me I would get a resulting purple point of 4." (Note that I am drawing the new points in purple.)

Together, we do the same thing at $x = 2$. "What do you guys think?" I ask. "What function are the purple points creating?"

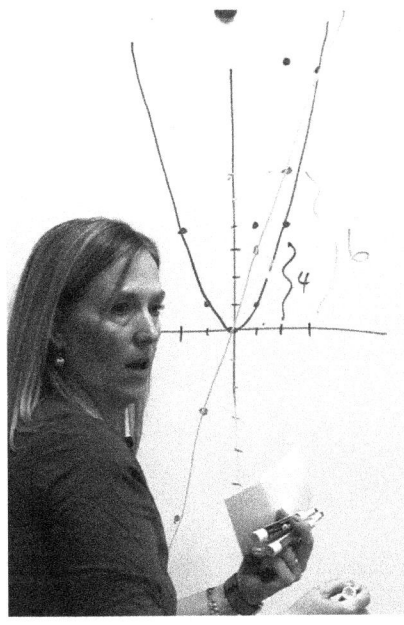

We then establish (0, 0) on the combined function because, as Ashlyn says, "The red one is also at 0." (Again, in real life I am using a red marker for the line.)

I connect those three purple points and ask, "Do you think it's [the combined function] going to be a line? . . . A parabola? . . . Maybe it's a cubic?"

As I point to the right, I ask, "Who could reason with me—what's going to happen further out?... Could we do the same thing for the x-value of 3? We could put in a value of 3 for each of them. What would happen? Is the combined function going to come down?" as I point down.

"Is it like going to do this?" as I motion the graph continuing to go up.

"Is it going to do this?" as I motion flatlining to the right.

"Maybe it'll do that?" as I motion that the graph would back up to the left.

"Like what's going to happen as x gets big?" as I point to the right. These are big ideas of end behavior—what happens in the extremes.

Ashlyn asks, "Will it just keep going?"

I respond, "Which way?"

Ashlyn answers, "Up."

I push for reasoning. "So we have a supposition here. We think, as x goes to the right, the graph is going to keep going up. Can you support that? Why?"

Ashlyn says, "Well, because the parabola and the linear are still going to keep going the same direction. They're still going to keep going out and up." She does not treat the graph as a static shape, a U that stops. She knows it will go *out* and *up*. She's treating the graph like the infinite relationship it is.

I press, "Anybody agree with that? Disagree with that? Can anybody add on to that? What else can you say about the quadratic and the linear, that they keep going up? What is their sum going to keep doing?"

Colby responds, "They infinitely go up, so you just keep going." He's reasoning about infinite sets.

Sarah suggests that if you add two positive y-values together their sum is positive. She's generalizing. So I draw the graph continuing to increase to the right.

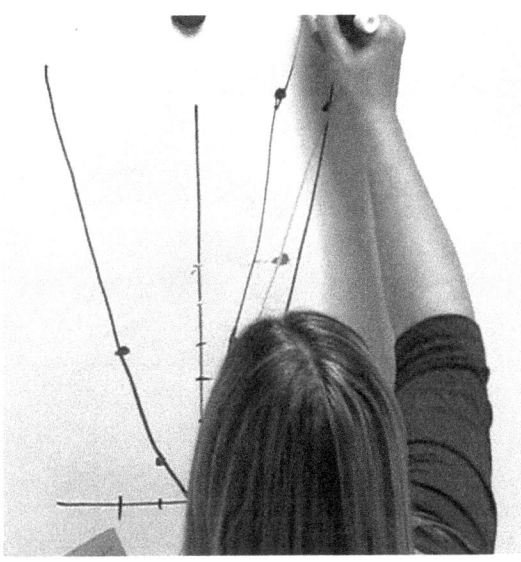

I then switch directions and point to the left. "Could we reason about what's happening back here? As x gets really small, negative? The functions are not both positive, so that's weird." To the left of zero, the x^2 is above the x-axis, but the $3x$ is below the x-axis.

To help, we quickly do a bit of point-by-point graphing in the second and third quadrants.

What's happening at $x = -1$?	What's happening at $x = -2$?	What's happening at $x = -3$?
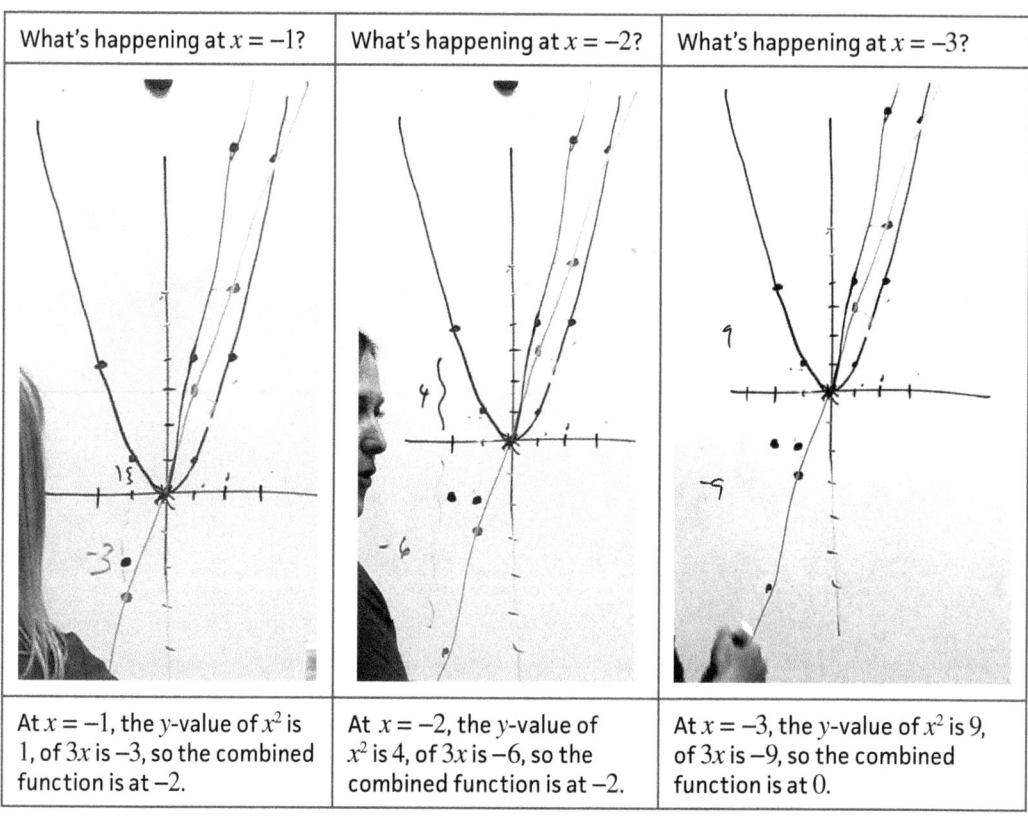		
At $x = -1$, the y-value of x^2 is 1, of $3x$ is -3, so the combined function is at -2.	At $x = -2$, the y-value of x^2 is 4, of $3x$ is -6, so the combined function is at -2.	At $x = -3$, the y-value of x^2 is 9, of $3x$ is -9, so the combined function is at 0.

Once we connect those points, Colby notices a connection. "I think it shifted. It shifted. Just the parabola." He treats the parabola as an object, a set of infinite points that can be shifted.

Devonte suggests that the parabola has shifted to the left and down.	Original $y = x^2$	Shifted left and down.

Next, we turn our attention to what would happen to the left of that zero at $x = -3$. "Over there," I say, pointing to the right, "we have positive plus a positive, but back here," pointing the left, "we've got...."

Colby says, "Now it's at the point where the positives are going to be bigger numbers than the negatives, so it's just going to stay positive."

I ask, "Do you agree with that? Can you repeat what he just said?"

Sarah responds, "Since it's on the left side of the graph, the positives are going to be larger than the negatives. When you add them, the sum is going to be a positive." Sarah is generalizing about values as x approaches negative infinity. Students are thinking about the relationship between squaring a negative number and multiplying that same number by 3 and combining those results. A lot happening simultaneously!

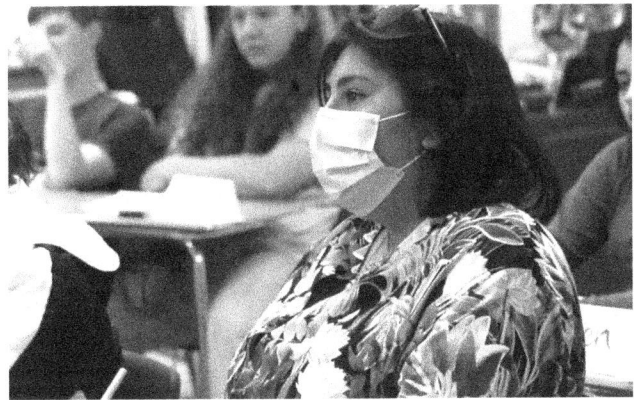

After a short discussion supporting that end behavior by looking at the values for $x = -1000$ to think about an example in the extreme, I ask the next question.

"What does this graph look like, $y = x(x + 3)$?"

Colby smiles and says, "I see what you're doing. You're trying to tie it all back together."

Ashlyn adds, "It's the same thing. It can be $x^2 + 3x$ because you're going to distribute the x."

Then the bell rings.

Are the graphs of the two functions the same, $y = x(x + 3)$ and $y = x^2 + 3x$? Does the factored form tell you anything different? What do we know about multiplying and getting a product of

Watch students develop functional reasoning

https://qrs.ly/ sog3rip

zero? Can you see those zeros at (–3, 0) and (0, 0) just popping out because of the zero product property? This is an example of reasoning about the short run behavior of that combination function.

To watch these students in action, see the QR code on this page.

In this experience, we see Functional Reasoning being built:

- Two variables, x and y, are linked and varying together, creating sets of ordered pairs (x, y).

- Two functions, $y = x^2$ and $y = 3x$, being treated as both a *process* (plug numbers in and get an output) and as *objects* (the terms x^2 and $3x$, and also a parabola and a line) and connections between multiple representations (equations, table values, graphs).

    - Considering individual points and considering the overall pattern of the sets of points simultaneously, including constant and changing rates of change.

    - Long run behavior as x approaches positive and negative infinity, a zoomed-out look (heading toward the idea of limits).

    - Short run behavior as x approaches zero from the left and right, a zoomed-in look (heading toward continuity, critical points).

TIP

When we refer to a function or to function values, we mean the y-values at certain x-values. If we ask, "Where is the function positive?" we mean, "At what x-values are the y-values positive?" Read back over the vignette to find places where I reinforced this notion. For example, when I pointed to the left and asked, "Could we reason about what's happening back here? As x gets really small, negative?" Clarifying that we are looking for small, negative xs to answer the question about what is happening with the ys can be helpful.

What is Functional Reasoning? The domain of Functional Reasoning means dealing with two linked variables that are varying in tandem. Whereas in Proportional Reasoning, two *quantities* are linked and varying in tandem, in Functional Reasoning, two *variables* are linked and varying in tandem.

"Functional Reasoning, also called Relational Reasoning or Bivariate Covariation, means that students consider the effect of the rate (which is a ratio—Proportional Reasoning) on the parent function, which is (often) a set of infinite points that follow a rule (which often contains additive and multiplicative relationships)" (IES, 2024).

To reason functionally, one must grapple with more and more simultaneously and in new ways.

"Because they are considering 2-dimensional points, this means that the student is also considering the effect the domain has on the range as it interacts with the rule. Many things are happening simultaneously and the learner has to focus on parts in the short run while also considering the whole in the long run" (IES, 2024).

Authors from the Calculus Consortium write that their approach to teaching about functions in algebra class "is to foster strategic competence and conceptual understanding in algebra, in addition to procedural fluency. . . . Strategic competence and conceptual understanding in algebra mean being able to read algebraic expressions and equations in real-life contexts, not just manipulate them, and being able to make choices of which form or which operation will best suit the context. They also mean being able to translate back and forth between symbolic representations and graphical, numerical, and verbal ones" (McCallum et al., 2010).

I call this domain *Functional* Reasoning, but it also applies to relations. This is not about function notation or a vertical line test; it's about sets of ordered pairs (x, y), and how those sets of ordered pairs are related and vary together.

The technical definition of a function is a set of ordered pairs (x, y) where for every x there is exactly one y. A relation is the more general collection of ordered pairs (x, y). Ideally, we would call this domain not Functional Reasoning but Relational Reasoning. Because the word relational is used colloquially to mean any kind of reasoning about relationships, that would be confusing! So, even though I am calling it Functional Reasoning, the reasoning is not only restricted to functions but also applies to relations.

FREQUENTLY ASKED QUESTIONS

Q: In the $y = x^2 + 3x$ Problem String, why didn't you sketch all that on a grid? That would have made the whole experience better, right?

A: I absolutely could have projected a grid onto the board or have done the whole String on graph paper on a tablet, etc. But I sketched the functions on a blank canvas on purpose. In fact, the whole *slightly less than exact* look of the graphs is purposeful. If I had given students graph paper and I was sketching on a grid, the chance of students getting stuck in counting grid lines is high (Counting Strategies). Since we were all on a blank canvas, it raises the potential for students to think about the quantities and the relationships between them, rather than counting. Also, the slightly messy appearance has the potential to nudge students to have to generalize because they cannot see exact points. That is not to say that there are not times when I deliberately put students and the display on a grid, but not for this experience.

COMPONENTS OF FUNCTIONAL REASONING

The graphs here show some components of Functional Reasoning.

Simultaneity

x	x^2	$2x$	$x^2 + 2x$
$-$	$+$	$-$	$+$
-2	4	-4	0
-1	1	-2	-1
0	0	0	0
1	1	2	3
2	4	4	8
$+$	$+$	$+$	$+$

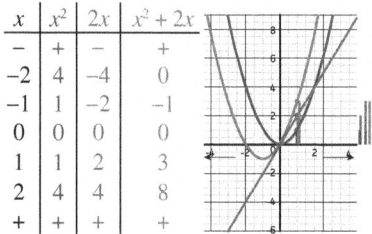

Dealing with more simultaneity, considering

- The representation as both a process and an object
- Both individual points and the overall pattern
- Connections between multiple representations
- The range depends on the domain

Short and Long Run Behavior

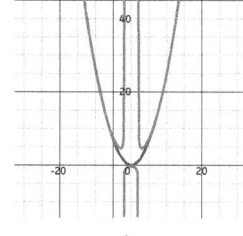

$$r(x) = \frac{0.25x^4 + 2}{x^2 - 4} \approx 0.25x^2$$

Considering both short and long run behavior

- What's happening near $x = 0$ (short run), zoomed-in look
- How is the function behaving as x gets really large (to the right) or as x gets really small (to the left), zoomed-out look?

Function Versus Equation

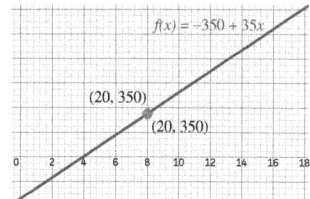

A "function" approach to teaching algebra includes considering simultaneously

- The function (or relation) represents the whole scenario
- Invest $350 into a lawn mower, make $35 per lawn mowed
- An equation, $350 = -350 + 35$ (20), represents one point on the graph $(20, 350)$, when you've mowed 20 lawns, you have made $350.

Source: Adapted from IES (2024).

Functional Reasoning is also shown in the thinking for problems like:

- Finding an appropriate viewing window for the T-shirt problem.
- Describing the relationship between the graphs of $y = x^2$ and $y = 2x^2$, reasoning about how specific transformations affect a parent function.

TRY IT

Find an appropriate viewing window for the graph of our T-shirt problem, where the revenue is the $200 setup fee, and charge $15 per shirt, $r(n) = -200 + 15n$.

The student reasons that n represents the number of shirts sold and decides that values from 0 to 30 shirts make sense. The student reasons that if they sell 0 shirts, they are in debt $200, and if they sell 30 shirts, they will have $-\$200 + \$15(30) = -\$200 + \$450 = \$250$, so a window [0, 30] for the independent variable means a window of [−200, 250] for the dependent variable. This is reasoning functionally because the student realizes the maximum revenue (output) depends on their chosen maximum number of shirts (input).

If a student randomly guesses the maximum value for revenue (output) *even though* they have already chosen the number of shirts (input), they are not reasoning functionally about that *yet*.

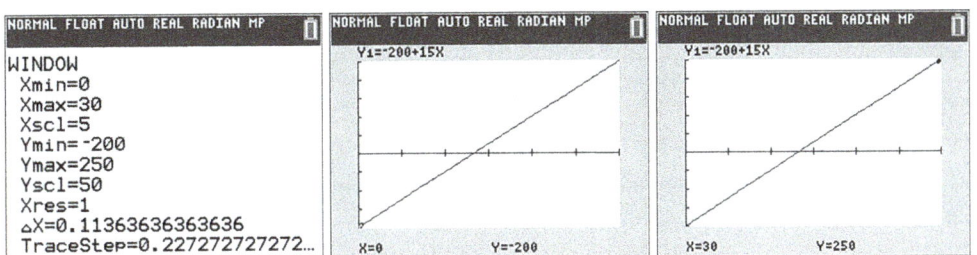

Source: Graphs generated by TI-84CX, Texas Instruments.

A student is reasoning functionally when they realize and use the idea that the range depends on the domain, the y-values depend on the x-values.

TIP

To maintain the benefits of finding appropriate viewing windows, refrain from using the zoom or automatic window setting functions. Instead, talk through what the variables represent and what values make sense for those variables. In our T-shirt problem, ask what the x represents in the situation (number of T-shirts) and what values make sense for the number of T-shirts (0 to 30 shirts). Once you've set those parameters, ask what the y-values represent (amount of money) and what values make sense for the amount of money given those number of shirts. If students guess at the minimum and maximum amounts of money, don't just *tell* them they should use their chosen x-values in their function. Instead, help them come to that conclusion by continuing to ask them to justify their choices. Also, listen for the student who uses their chosen x max to decide the y max. Wonder aloud, "Why would you choose that value?" Ask them to defend their choice. Don't require it too soon. Let it arise in a few Rich Tasks.

TRY IT

Describe the relationship between the graphs of $y = x^2$ and $y = 2x^2$.

A student demonstrates Functional Reasoning when they describe a parabola as an infinite relationship between numbers and their squares. A student also demonstrates Functional Reasoning when they describe the transformation as the y-values are getting bigger faster for a function that continues to increase as x gets larger to the right and

(Continued)

(Continued)

smaller to the left. For each increment in x-values to the left and right, the y-values are twice as high. For example, in $y = x^2$ we have the point $(1, 1)$ but for $y = 2x^2$ at that same x-value, the y-value is twice as big $(1, 2)$. The same thing happens for $(4, 16)$ to $(4, 32)$. The y-values get bigger faster. Similar Functional Reasoning describes the transformation of $y = x^2$ to $y = \frac{1}{2}x^2$ as the y-values getting big half as fast or slower by $\frac{1}{2}$.

A student does not show Functional Reasoning when they describe a parabola as a U shape where $y = 2x^2$ is skinnier, narrower, or smaller than $y = x^2$. First, describing the letter U implies that a parabola is a static, finite shape, not the graph of infinite quadratic relationships. Second, the words *skinnier, narrower,* and *smaller* imply that scaling the parent function by 2 made the function values (y-values) lower when, in fact, it made them higher. Similarly, students are not reasoning functionally when they describe the relationship between the graphs of $y = x^2$ and $y = \frac{1}{2}x^2$ as "wider, fatter, bigger," which implies that scaling the parent function by $\frac{1}{2}$ made the function values greater, when in reality it made them less.

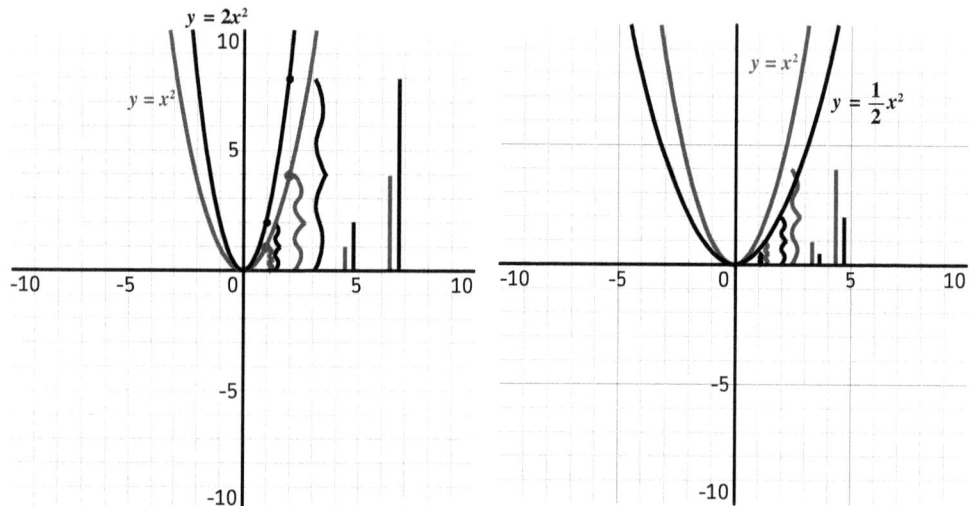

A student is reasoning functionally when they realize and use the notion that polynomials are infinite relationships, that the graphs do not end. The graphs are not static shapes.

A student is reasoning functionally when they describe the effect that scaling a function has on the y-values (often referred to as the function values), not just describe the look of a graph. Realizing the perspective of referring to *what the function is doing* is the y-values and *where the function is doing it* is the x-values.

DMR Online
Workshop

https://qrs.ly/
6rg3rek

While there are many procedures and algorithms happening in high school that can trap students into using less-sophisticated reasoning, we will focus on the algorithms associated with writing the equation of a line.

Video example
of Functional
Reasoning

https://qrs.ly/
qng3riv

WRITING THE EQUATION OF A LINE

How did you learn to write the equation of a line when given points?

Join me in Abby's Sanchez's ninth-grade algebra 1 classroom. Abby starts the class period in the second week of the year with a very deliberate question, "How does your walk affect the graph?" She gives each group of three students a motion detector connected to a graphing calculator so they can experiment and see the results of their walks. The calculator plots elapsed time in seconds as the x-value and the distance from the sensor in feet as the y-value.

Video non-example
of Functional
Reasoning

https://qrs.ly/
ypg3riw

In small groups, students walk in front of motion detectors, comparing their walk to the resulting graphs, noting their observations, and, with Abby's skilled questioning, refining their ideas.

Video example
#2 of Functional
Reasoning

https://qrs.ly/
k5g3riz

The next day, Abby guides a class discussion to generalize three important ways their walk affects the graph:

1. Direction—Walking away from the motion detector makes the graph go up, walking toward the motion detector makes the graph goes down, from left to right.

2. Speed—Walking faster makes the graph steeper, walking slower makes the graph less steep, from left to right.

3. Starting point—Where you start walking affects the *beginning* point of the graph at time 0 seconds, which is the *y*-intercept.

Video non-example
#2 of Functional
Reasoning
https://qrs.ly/
sgg3rj1

Abby follows these two days with several days of more tasks and Problem Strings to continue to build these ideas in different contexts like saving and spending money, earning money at a lemonade stand, driving vehicles away and toward a city at different rates.

Then, Abby brings the motion detectors back out.

"Remember when we walked in front of these gizmos?" she asks. "What if someone started 2 feet in front of the sensor and walked away from the sensor at 1.5 feet per second? What would that walk look like? That graph look like?"

The students and Abby together find a few points and they plot them.

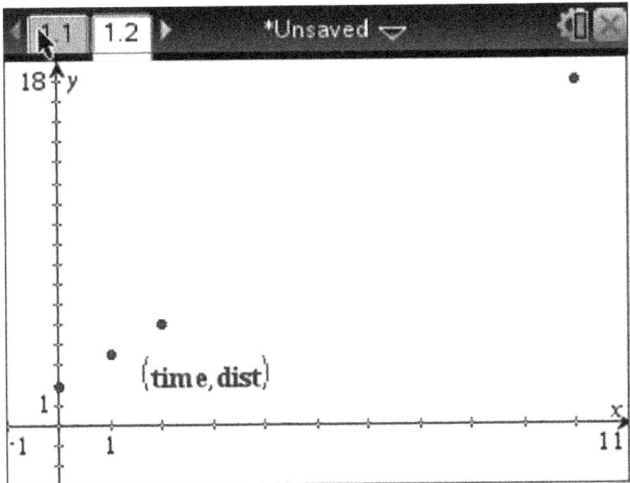

Source: Graphs generated by TI-NSpire, Texas Instruments.

"What could you say about the position of the walker at time t?" asks Abby. With her questioning and guidance, students say: "The distance at time t equals start at 2 feet and walk away at 1.5 feet per second" as Abby writes $D(t) = 2 + 1.5t$, modeling their thinking with that function rule. This is not Abby telling them a rule. She is putting symbols to what the students are telling her about the context. They add the graph of that line to the scatter plot.

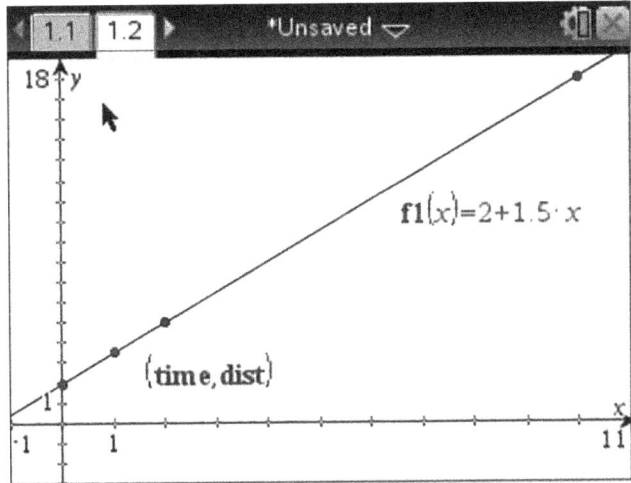

Source: Graphs generated by TI-NSpire, Texas Instruments.

"Sure looks like we wrote a function that represents that walk," Abby says. "What if we actually walk in front of the motion detector? Could we write a function for those points?"

With the instructions to walk away at a constant slow rate, a student walks and they look at the graph.

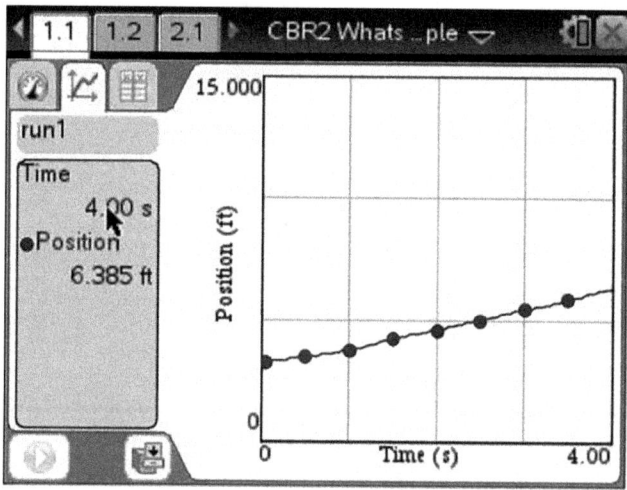

Source: Graphs generated by TI-NSpire, Texas Instruments.

"How can we find the starting point and rate for this walk?" Abby asks.

Together, students decide to use the calculator's trace function to find points on the graph.

They find the first data point collected (0.05, 3.34), so Juan started around 3.34 feet in front of the sensor.	They trace to 1 second, so at 1 second, Juan was 3.81 feet in front of the sensor (1, 3.81).

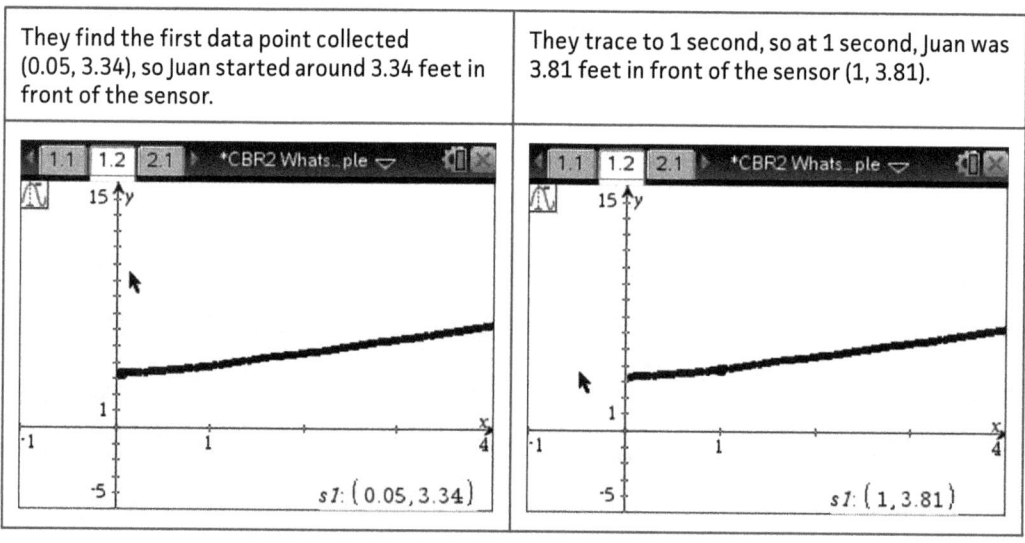

Source: Graphs generated by TI-NSpire, Texas Instruments.

Source: Graphs generated by TI-NSpire, Texas Instruments.

Abby asks, "How far did Juan go in 1 second?"

When the students answer "0.47," Abby asks, "So Juan walked 0.47 feet in 1 second?"

They enter the function, distance $f(x)$ equals start at 3.3 and walk away at 0.47 feet per second times the number of seconds, x.

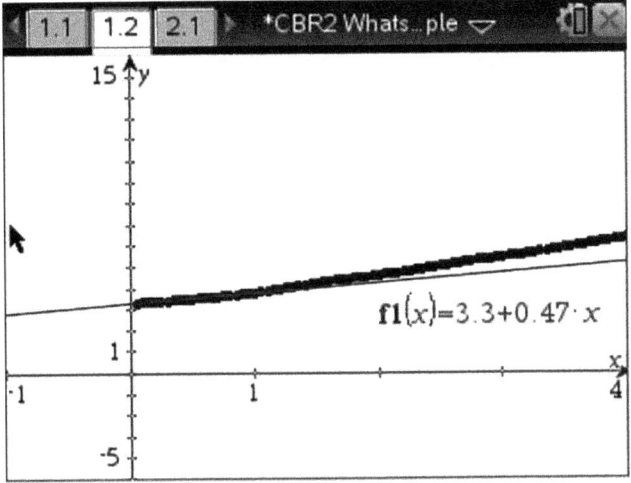

Source: Graphs generated by TI-NSpire, Texas Instruments.

Students note that Juan sped up! He walked faster than 0.47 feet per second during the last few 1-second intervals.

The class then does several more walks, each time making sense of how the function they write relates to the way the walker walks.

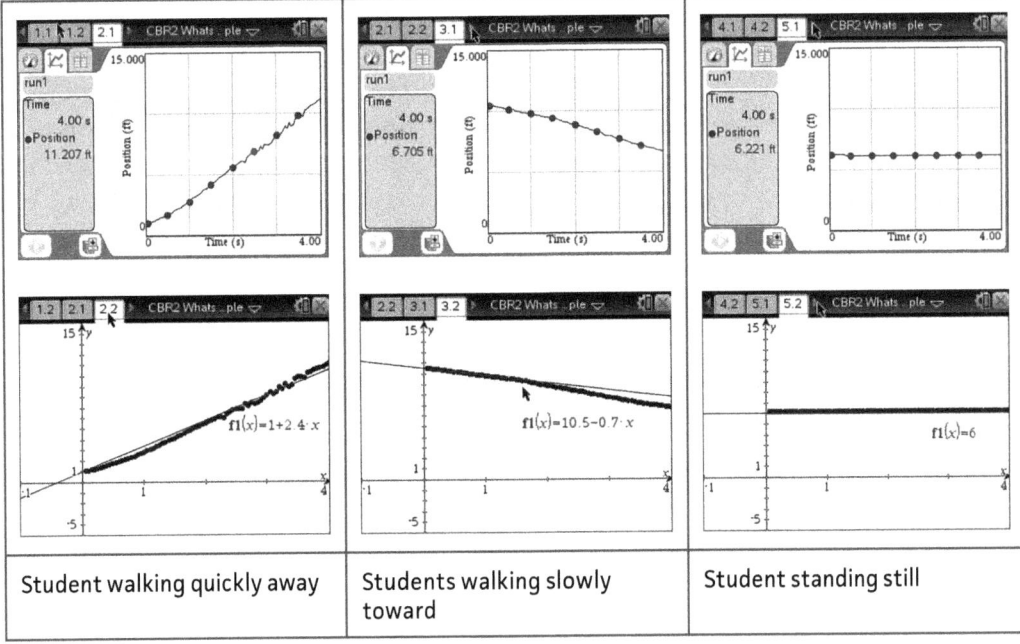

Source: Graphs generated by TI-NSpire, Texas Instruments.

After working through several different walks and writing equations to match the walk, Abby notes, "It's almost like every time we wrote the function to match the walk, we thought about the *distance* at time *t* equals how we *began* plus how we *moved* times the *time*," as she writes $D(t) = b + mt$.

These students are building Functional Reasoning as they grapple with and make sense of ordered pairs, distance versus time graphs, starting points (*y*-intercept), slow and fast rates, direction (away and toward), and how all of that can be represented as a function rule with variables.

Abby's classroom shows us what it can look, sound, and feel like to learn to reason about linear functions.

TIP

A motion detector can be used to gather other helpful data, like the height versus time for a bouncing ball where you extract the heights of the bounces and graph (bounce number, height) for a look at exponential decay. You can study velocity and acceleration for constant and nonconstant rate walks. You can even drop a very light object from above the motion detector to study gravity and projectile motion.

Read the next section to find steps that show how students can get trapped into writing the correct equation of a line but never building Functional Reasoning about lines.

THE TRAP OF TRADITIONAL ALGORITHMS FOR WRITING EQUATIONS OF LINES

In the United States, textbooks have traditionally suggested a set of formulas for writing the equation of a line. Typically, students are directed to determine what information is given, decide which formula(s) to use, plug in values, and *simplify* the resulting equation. There is nothing wrong with the formulas themselves—they correctly represent the phenomena. I am suggesting that it is problematic to dictate a series of steps for students to mimic.

Another use of the word simplify. *This time it means to distribute and collect like terms. See Chapter 5, where* simplify *meant something completely different with fractions.*

These steps seem straightforward enough, yet they set sneaky traps for learners who should be developing Functional Reasoning with linear functions.

Let's walk through an example problem, considering how a student could find the correct equation of a line while being trapped by this procedure to never reason at all about rate of change, constant rates, slope, distance versus elapsed time, or the graphic linear representation.

Example problem: Write the equation of the line containing (2, 7) and (6, 19).

To begin, the student determines what information is given and chooses from a menu of formulas.

STEP 1 OF WRITING THE EQUATION OF A LINE (OR CHOOSING THE CORRECT FORMULA)

Students choose from one of these formulas:

$y = mx + b$	$y - y_1 = m(x - x_1)$	$Ax + By = C$	$m = \dfrac{y_2 - y_1}{x_2 - x_1}$
Slope y-intercept formula	Point slope formula	Standard form	Slope formula

The student determines they've been given two points, so they'll use the two-point formula. Wait, there is not a two-point formula. Why isn't there a two-point formula? Sigh. So, the student decides to find the slope using the slope formula and then use one of the points and that slope in the point-slope formula.

STEP 1: FALSE DEFINITION OF MATH TRAP

This "determine what is given and choose from a list of formulas" way of thinking builds recipe-followers but not chefs.

FREQUENTLY ASKED QUESTIONS

Q: But don't we need to teach students to follow a recipe? Isn't it a good skill to be able to ascertain which formula is appropriate and then successfully follow the formula? And isn't math class the perfect place to teach this skill?

A: Even if we determine that it is a desirable skill to be able to choose an appropriate formula and successfully follow steps to get answers, mathematics class is not the place to build that skill. Why?

(Continued)

(Continued)

Because following a recipe is not math-ing. This is the crux of my argument—the purpose of math class is to build mathematical reasoning, not step-following. I believe you when you report that your experience as a student was learning to follow steps and that success was defined as choosing the correct procedure and mimicking that procedure correctly. What if success in mathematizing is actually something quite different? By now, you know: It *is*.

Everything from here on will go awry if students initially choose an incorrect formula.

This happened once to me. In my 11th-grade Math Analysis (precalculus) class, I learned to graph quadratic functions in a very algorithmic way. I had it down to a tee, every step exactly just right. On the day of the test, I dutifully solved the problems, meticulously showing each and every step.

When I got my test back, my heart sank. A 50%?! That was a new experience for me. What had happened? I had graphed each version of a $y = x^2$ as a left-right opening parabola and every $x = y^2$ as a vertical up-down opening parabola. Every single step after that was correct. Yet, in actuality, every one of the problems was entirely wrong. I was supposed to be baking a cake, but the recipe for bread did not produce a cake, even if I did every bread step correctly. Not really understanding what I was doing, I had chosen the wrong recipe for each problem.

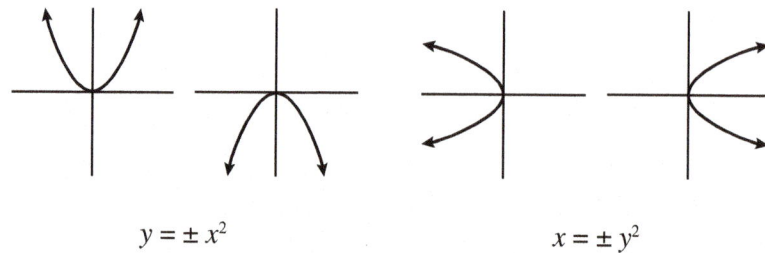

$$y = \pm x^2 \qquad\qquad x = \pm y^2$$

STEP 2: FIND THE RATE OF CHANGE (SLOPE)

If the student has made it to this step by choosing a correct formula to find the rate of change (slope) for the line, the student plugs in the values and computes.

For $m = \dfrac{y_2 - y_1}{x_2 - x_1}$ and (2, 7) and (6, 19), $m = \dfrac{19 - 7}{6 - 2}$.

STEP 2: THE PROCEDURAL THING-TO-DO TRAP OF THE SLOPE FORMULA

If students approach the slope formula as a *thing to do*, chances are very high that students will approach the formula as *minus, minus, divide.*

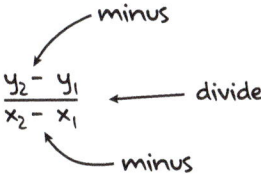

Because slope is a *thing to do* and not an overarching, important idea to grapple with and make sense of, students do not grapple with and make sense of constant rates of change.

STEP 2: TRAPPED IN REMOVAL SUBTRACTION

Also, if students are in the procedural minus-minus-divide trap, this means using solely the removal (minus) meaning of subtraction. Remove in the numerator, repeat for the denominator, and then divide the results.

For $m = \frac{y_2 - y_1}{x_2 - x_1}$ with (2, 7) and (6, 19), $m = \frac{19 - 7}{6 - 2} = \frac{12}{4} = 3.$

Remember that subtraction has two important meanings: removal *and* distance/difference (see Chapter 3). Being stuck in the *thing to do* trap means that students use the *removal* meaning of subtraction. This is unfortunate because the subtractions in the slope formula have nothing to do with removal (minus) and everything to do with distance/difference. What would it even mean to remove a y-value of a point from the y-value of another point? But it makes all sorts of sense to find the distance between the y-values of those points. Ditto with the x-values.

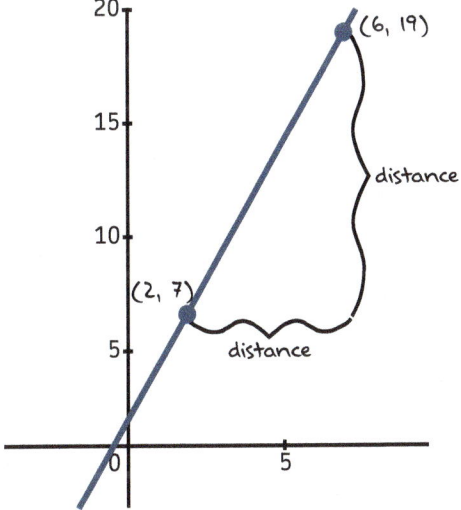

What is a rate of change, anyway? In this case, what is a rate of change of 3? It is actually a rate of 3 to 1. A rate is a ratio of two quantities, like speed (miles per hour, feet per second), concentration (scoops of mix per gallon), interest rates (percentage rates for a loan). Rates are Proportional Reasoning! Ideally, students would have done earlier work to really reason about rates. Even if they had, this algorithmic approach to finding slope can trap students into not thinking about the ratio of the distances (magnitude) with direction (sign of the rate).

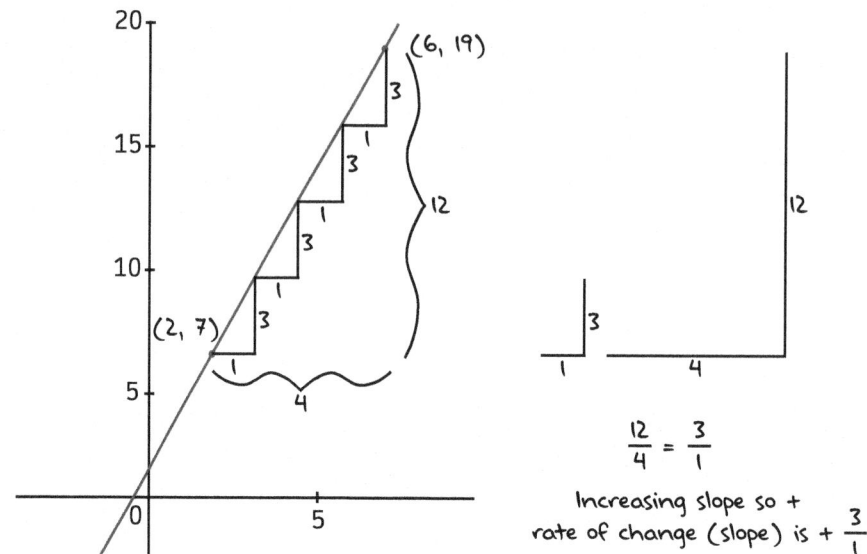

$$\frac{12}{4} = \frac{3}{1}$$

Increasing slope so +
rate of change (slope) is $+\frac{3}{1}$

When students solve problems with integers in a rote-memorizing mimicking way, they remember: keep, change, change (or is it keep, change, flip?).

For (–2, 10) and (1, –2), $m = \dfrac{10-(-2)}{-2-1} = \dfrac{\textit{keep change change?}}{\textit{keep change flip?}}$

Instead, we can reason about the ratio of distances (magnitude) with direction (sign of the rate). The following is an example with points that indicate a negative slope. Students can reason about the ratio of the vertical and horizontal distances and then reason about increasing or decreasing slope.

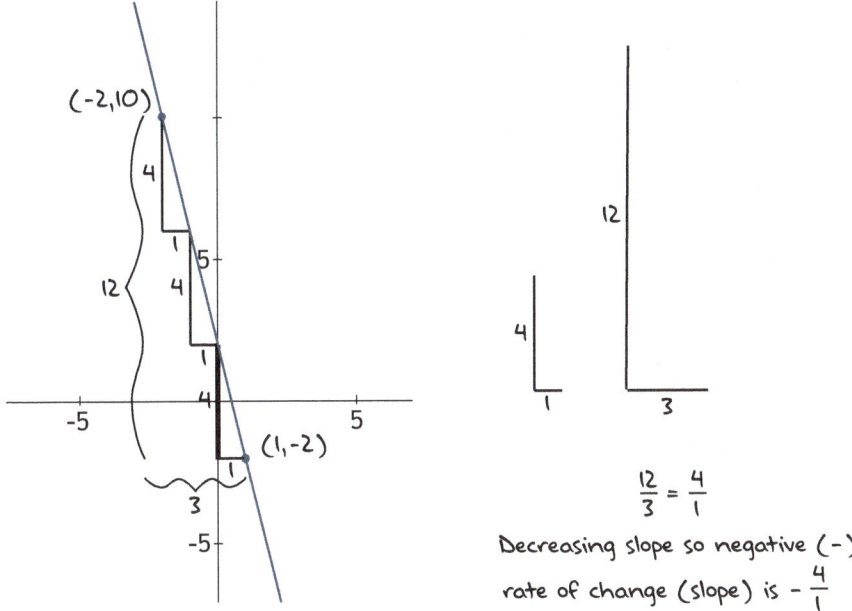

$$\frac{12}{3} = \frac{4}{1}$$

Decreasing slope so negative (-)

rate of change (slope) is $-\frac{4}{1}$

If students do this *minus, minus, divide* series of steps, they could be doing that subtraction with Counting Strategies (see Chapter 3 for all of these traps). If students are using the long division algorithm to divide, they could be either using no mathematical reasoning with mnemonic devices, Additive Reasoning, or at best single-digit Multiplicative Reasoning (see Chapter 4 for all of these traps). All of these less-sophisticated methods do not build Proportional or Functional Reasoning.

STEP 3: PLUG AND CHUG

Now that students have found the rate, students substitute the rate and a point into the formula. If they have chosen a *correct* formula in step 1, for (2, 7) and (6, 19) and the slope of 3:1 and $(y - y_1) = m(x - x_1)$, they could now have either:

$(y - 7) = 3(x - 2)$ if they use the point (2, 7) or $(y - 19) = 3(x - 6)$ if they use the point (6, 19).

STEP 3: THE TRAP OF ROTELY USING FORMULAS YOU COULD BE UNDERSTANDING

Because so many of us were taught this point-slope form of a line as a formula to *do*, we never had the chance to connect it to related big ideas: in this case, transformations of functions.

This "formula" would be far better off in a slightly different form and delayed until after students have constructed transformations of functions and realized that $y = m(x - x_1) + y_1$ is simply the line $y = x$ being vertically stretched by a scale factor of 3:1 and shifted to the point (x_1, y_1).

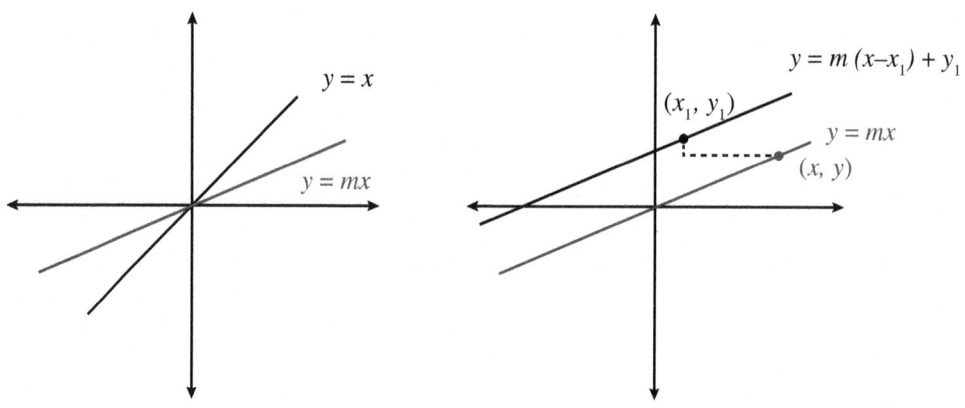

In this case, the line is scaled by 3 and shifted right 2 and up 7. This is consistent with transformations of all functions.

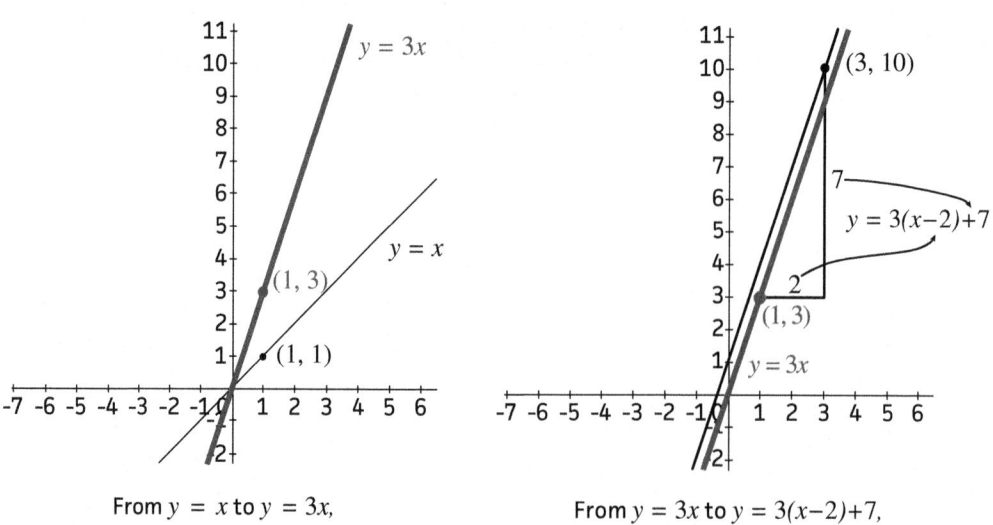

From $y = x$ to $y = 3x$,
y-values are scaled by 3.

From $y = 3x$ to $y = 3(x-2)+7$,
the line shifts right 2 and up 7.

But rather than build the point-slope form of a line based on transformations of functions, teachers and students are trapped into a formula "do stuff" mentality.

TRY IT

What could it look like to reason functionally to find the equation of the line containing (2, 7) and (6, 19)?

- Consider the two points as locations of a walker, walking in front of a motion detector. How could that help you find the rate of the walker?

- How could you use the rate to find the starting point of the walker?

- How could you use the rate and the starting point to find the function that represents the walk?

How could you use the points to represent a walk in front of the motion detector to find the rate between (2, 7) and (6, 19)? Those data points can represent: at an elapsed time of 2 seconds, the walker was 7 feet in front of the motion detector, and at an elapsed time of 6 seconds, the walker was 19 feet in front of the motion detector.

You can find the rate of change (slope) by reasoning that in 4 seconds, the walker traveled 12 feet—that's 12 feet in 4 seconds, that's 3 feet per second. You can visualize the points to determine that the line is increasing and therefore the rate is positive.

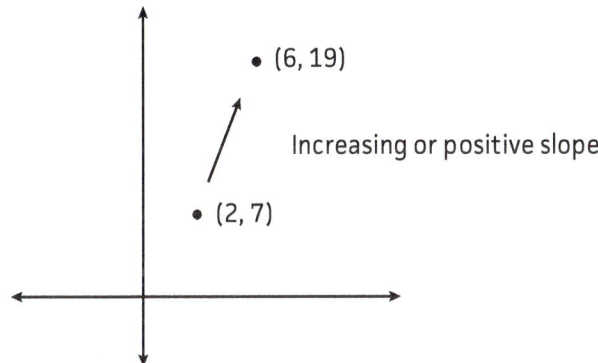

How could you use the rate to find the starting point of the walker?

The point (2, 7) means that the person was 7 feet away from the motion detector at 2 seconds. Since the person was walking at 3 feet per second, in those 2 seconds the person traveled 3 feet per second times 2 seconds, or 6 feet. If the person traveled 6 feet and landed at 7 feet from the motion detector, they must have started 1 foot in front of the sensor.

(Continued)

(Continued)

How could you use the rate and the starting point to find the function that represents the walk?

Since the distance at time t is where you began plus how you move times the time t, $D(t) = b + mt$, the distance at time t is start at 1 foot in front of the sensor and walk away at 3 feet per second times t the number of seconds: $D(t) = 1 + 3t$.

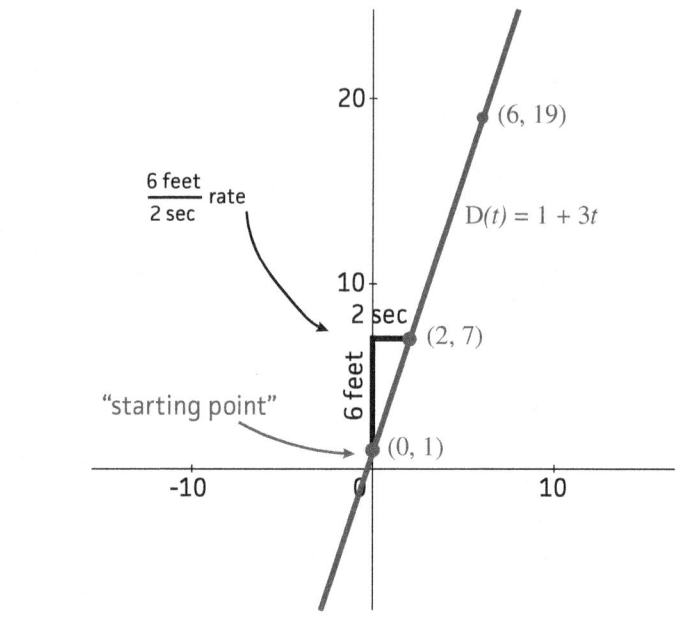

This is developing and using Functional Reasoning: building on understanding of rates; motion graphs; the linear parent function, $y = x$; and the functions students write for walking away and toward the motion detector at constant rates.

Conclusion

"Quit making me think!" Rebecca cries as she throws a graphing calculator at me. It is more than a toss but not quite a fastball pitch.

Surprised, I catch the calculator and ask her what is wrong. Rebecca has newly transferred into my high school precalculus class during the semester break. She had been a student in my German 2 and 3 classes and decided to switch to me for math as well. A few days into the second semester, she is frustrated.

"I do *not* understand what is going on in precalc," she says. "It's not like any math class I've had before."

All year long, the students in my precalculus classes have been math-ing, reasoning, making sense of, and using relationships. Rebecca has been extremely successful in her earlier math classes, rote-memorizing and mimicking with the best of them. Because this class is so different, she is seriously off-balance. Since we already know each other in German class, it hasn't occurred to me to help her acclimate to my math class.

"Ah, right. I get it. You're used to being told what to do and repeating that, right?" I ask.

"Yes, that's what math is," she answers with some emotion.

"Math is actually way cooler than that, and you're going to be spectacular at it!" I say, and then we sit and talk through expectations and help her get her bearings.

Once she relaxes into thinking and reasoning, she does indeed succeed spectacularly that year and the next in calculus. I've lost track of Rebecca over the years since then, but I haven't forgotten her discomfort at finding herself in an unfamiliar place, so different from what she was used to.

Perhaps you're finding yourself in a similar situation. Math is math . . . isn't it?

I invite you to relax into thinking, reasoning, and math-ing and have a spectacular time doing it!

This chapter is designed to provide insight into some of the big facets of Functional Reasoning. Reasoning about functions, relations, graphs, tables, equations, transformations, and more means dealing with many deeply interconnected concepts, models, and strategies that are all interacting simultaneously. Remember that grappling is necessary as you build your brain's ability and capacity to reason about increasingly complex mathematical relationships.

Discussion Questions

1. How did you think about functions and relations as a student? How has your experience influenced the way you have taught?

2. For the T-shirt problem and the graphing a rational function problem:

    a. How do you reason through them?

    b. Predict how your students are reasoning about these problems.

    c. Ask students an appropriate problem or two. How did they respond?

    d. How might you make your or your students' thinking visible?

3. How are your students reasoning about linear situations? Using formulas, in which they are using Counting Strategies or memorized rules to compute with integers to find the slope? Using Additive Reasoning as they reason recursively to find the sequence of y-values? Reasoning proportionally about the slope as a rate? To help you decide, you could use the "How are you thinking about linear situations right now?" questions on page 183.

4. How are your students reasoning about rational functions? To help you decide, you could use the "How are you thinking about rational functions right now?" questions on page 185.

5. What does it mean to think of a function as both a process and an object?

6. How has your Functional Reasoning developed while reading this chapter?

7. How have your ideas about developing Functional Reasoning with your students changed while reading this chapter?

TRY IT IN YOUR CLASSROOM

Find an Appropriate Viewing Window

"Finding an appropriate viewing window" does not appear in any set of course requirements. But using technology to search for and adjust viewing windows can help build important underlying relationships in an environment with low risk and low cost of failure.

Domain, range, parent function behavior, continuity, transformations, and more all come together in this brilliant routine. As students adjust the window, they see the function again and again, stretching, shrinking, reflecting, and shifting based on their vantage point.

Students continually witness the effect of transformations on those parent functions, the function $f(x)$ actually shifts to the right when x has been replaced with $x - 2$, $f(x - 2)$, and the whole function $g(x)$ actually shifts down 3 when 3 is subtracted from the whole function $g(x) - 3$.

Students get a chance to wonder why scaling a quadratic function $h(x)$ by 3, $h(x) = 3x^2$, leaves the *anchor* point $(0, 0)$ at the origin, but scaling an exponential function $g(x)$ by 3, $g(x) = 3 \cdot 2^x$, appears to shift its anchor point from (0, 1) to (0, 3).

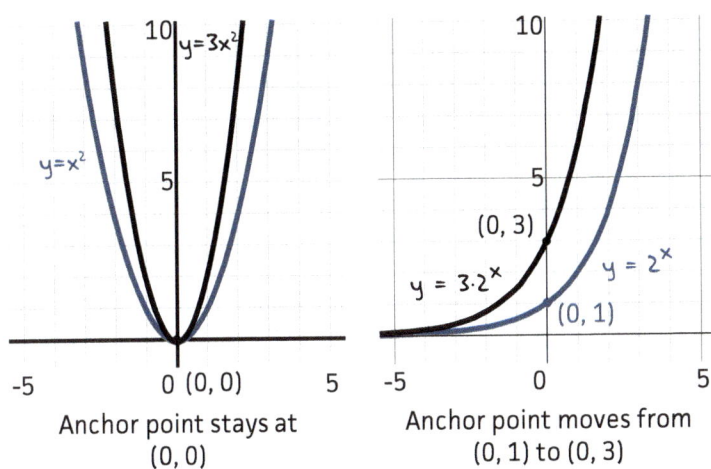

Anchor point stays at   Anchor point moves from
(0, 0)   (0, 1) to (0, 3)

And when scaling that square root function, it appears to be stretching vertically up, but scaling a trigonometric function results in it scaling vertically up *and* down?

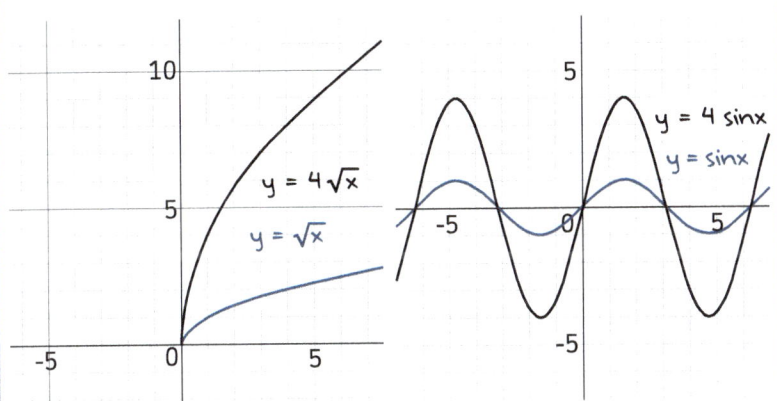

Purpose

Use technology to develop domain, range, parent functions, and transformations. Students also build the idea that the input/domain affects the output/range.

(Continued)

(Continued)

The Routine

- Give students a purposefully transformed parent function.

- Instruct students to find an *appropriate* viewing window. You can use your favorite graphing calculator, handheld or online.

  - Would a square window be helpful?

  - Can you see all of the important features of the graph?

- Tell students that the "zoom" or automatic window finding keys are broken or off limits.

- Have a few students share their chosen window, display them for everyone to see, compare and have them defend their choices.

Things to Consider

Choose parameters for the functions that place the function far enough out of a beginning window so that students cannot just zoom in or out. Try for functions that look odd to begin with, like $y = 100\sin x$ or $y = 100\sqrt{x}$. Also, use functions where the first look is deceiving in that it only shows part of the function, like a cubic that only shows the dip or only shows the increasing curve.

Questions to ask:

- How did you know where to look? What about the function rule helped you to know?

- What is the effect on the graph of adding/subtracting to/from the function, $f(x) \pm b$?

- What is the effect on the graph of replacing x with $x \pm a, f(x \pm a)$?

- What is the effect on the graph of scaling the function by a, $af(x)$?

- What is the effect on the graph of replacing x with $-x, f(-x)$?

- How can you use what you know about transformations to find an appropriate window?

Sequencing

Use a different parent function to work on different transformations (Murdoch et al., 2014; Murdoch et al., 2017):

- Parabolas $y = x^2$ work well for translations, like $y = (x + 175)^2$ or $y = x^2 - 350$

- The square root function $y = \sqrt{x}$ works well for reflections, like $y = \sqrt{-x}$ or $y = -\sqrt{x}$

- The absolute value function $y = |x|$ works well for dilations, like $y = |0.01x|$ or $y = -300|x|$

- The sinusoids, $y = \sin x$, $y = \cos x$, work well for periods, like $y = \sin 25x$ or $y = \cos 0.1x$

Then combine transformations, for example:

- $y = 200(x - 250)^2$

- $y = -(x - 300)^4 - 350$

- $y = \sin 50x + 25$

- $y = 0.1(x + 150)^3 - 175$

- $y = 30.2^x + 100$

Extension

Have students create their own, seeking for first looks that are odd or deceiving/incomplete. Students can trade with each other, seeking for clever functions and for ways of finding windows efficiently.

I give credit to Demana and Waits (1990) for this brilliant window finding idea from their textbook *PreCalculus Mathematics: A Graphic Approach*.

CHAPTER 7

If Not Algorithms, Then What?

I f teaching algorithms sets all these traps, then what are we to teach?

Meet Jordan, a typical second-grade student. His teacher, Stephanie Lugo, takes video of each of her students at the beginning of the year and then again later in the year to document how they are learning often-missed math facts.

In the video, we hear Stephanie ask, "Jordan, what is 15 minus 7?"

Jordan pauses, moving his eyes in an arc as if jumping on a number line in his brain, and says, "It's 8."

"How did you know that?" asks Stephanie.

"Because if you back 5, that'll be 10. And then you minus 2 that'll be 7, so it will be 8."

His strategy could look like this on an open number line:

Jordan knows that 15 – 7 is figure-out-able and that it's his job to figure it out. He doesn't have it automatically, *yet*. What he does have is confidence. He's math-ing.

To see Jordan mathing, watch the video at the QR code on this page.

Jordan was a student at the same elementary my kids attended. My oldest son also had Stephanie for second grade a few years earlier. These are the teachers I was working with and learning from.

Jordan had Kim Montague as his teacher in third and fourth grade. In fourth grade, I shot video of Kim's students answering some often-missed multiplication facts.

Watch 2nd-grade Jordan math-ing with subtraction

https://qrs.ly/ 7zg3rj2

"Jordan, what's 11 times 12?"

He pauses, taps his finger a couple times, and writes.

"$120 + 12 = 132$."

"How did you do that?" I ask.

"I knew 10 times 12 equals 120. And, um, then I just add 12 because 1 times 12 is 12. And I just add 12 to 120 and I got 132." During that explanation, Jordan never once looks down. He confidently goes about figuring out what he knows is figure-out-able.

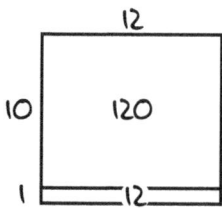

Now, there are other students in the class who just know 11×12. When I ask Brianna, "How do you know?" she smiles and replies, "We work with 12s a lot."

Jordan is a thoughtful student. He takes time to really think about things. He needs more experiences than some students. But given time and experience, he can mathematize, using what he knows to solve problems, building his brain in the process.

To watch Jordan in action, see the QR code on this page.

Like Jordan, mathematics teachers also must have the experience and confidence that math is figure-out-able.

Watch 4th-grade Jordan math-ing with multiplication

https://qrs.ly/w1g3rj3

STUDENTS AND TEACHERS NEED TO EXPERIENCE MATHEMATIZING

What does it look like to give the needed experiences to students and teachers?

In far too many places the conventional wisdom is to develop conceptual understanding so that students will choose the correct algorithm, perform it correctly and with correct answers, and then we have achieved the goal of mathematics education.

I'm telling you that the goal of mathematics education is to develop mathematical reasoning. This is not just a fluffy "think better" idea; it includes content. Learning the content is *important*. In fact, reasoning mathematically means owning content, deep down, inside and out, with a web of interconnected relationships.

So, if not algorithms, then what?

Just as students are in multiple places along the landscape of learning mathematics (Fosnot & Dolk, 2001), teachers are also in different places along the landscape of teaching mathematics. Just as some students need more experiences with some landmarks of mathematics, some teachers need more experiences with some landmarks of mathematics for teaching.

The following sections list major landmark stages on the teaching landscape we use at Math Is Figure-Out-Able to help teachers develop. As you read, decide where you might benefit from more experiences.

SOLVE PROBLEMS USING WHAT YOU KNOW

We all have more to learn in certain areas of mathematics. To understand what we mean by this stage, choose an area of mathematics that you teach. Before moving on to learn to teach it better, ask yourself:

- Do I rely on algorithms or procedures that someone showed me?

- Can I solve problems in this area using what I know, logically reasoning from one relationship to another, making sense as I go? Is this connected to my larger body of math knowledge, or does it exist as an island in its own little corner?

- When I'm reasoning, am I using the highest level of reasoning called for with these problems? For example, if it's an addition problem, am I counting by ones (less sophisticated reasoning), or am I using bigger jumps than one (Additive Reasoning), or am I tweaking the problem to make an equivalent one that's easier to solve (more sophisticated Additive Reasoning)? If it's a multiplication problem, am I skip-counting (using Additive Reasoning), single-digit Multiplicative Reasoning, or Multiplicative Reasoning with larger numbers (more sophisticated)?

If your answer to these questions is that you're relying on an algorithm or using less sophisticated reasoning than the problems call for, take heart! That's where I was not too long ago.

The journey to learn to solve problems using what *you* know is a fun and confidence-building process.

And there's good news! You won't have to figure it all out on your own.

TRY IT

Every chance you get, work to solve problems logically, starting from what you know and building from there.

Interact with others solving the same problems. Engage in #MathStratChat (see the end of this chapter) by solving the problems, reading and analyzing what others have tried, and commenting on their strategies. Focus on understanding, ask questions, and don't worry if you're not using a certain level of reasoning at first.

After you solve the problem, step back and look at the problem and the relationships you used again. Seek new insights on how you

---

might have used bigger jumps or bigger chunks, go a little over, scale by different factor, or how you might use the same connections in a similar problem.

Once you are solving problems in that area of mathematics using what you know, it's time to broaden your content knowledge and learn mathematics for teaching.

LEARN MATHEMATICS FOR TEACHING

Mathematicians, scientists, and engineers must all have mathematical knowledge and skill, but that does not mean they know the mathematics content for teaching. What's the difference?

It's one thing to be able to solve problems using what you know and it's another thing to know the major models, strategies, and concepts to teach that area of mathematics (Ball et al., 2008). Teachers need to know mathematics for teaching.

Part of the mathematics for teaching are the hierarchical progressions outlined in Chapter 2 (see Figure 7.1), how each domain relates to the next, how to recognize the reasoning being developed and the reasoning that students are using.

FIGURE 7.1 ● The Full Spectrum of Mathematical Reasoning With Longitudinal Domains

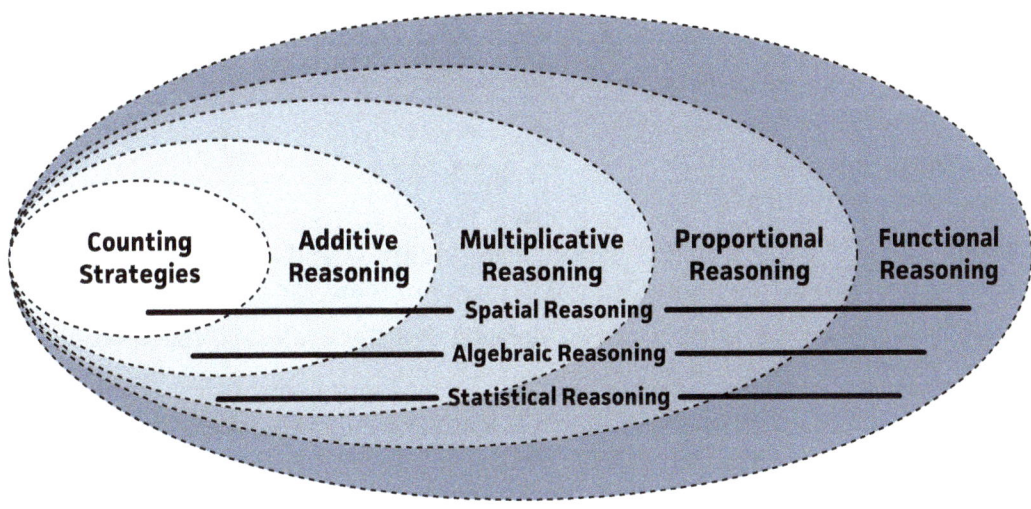

Here are four examples of important mathematical content for teaching. They do not make an exhaustive list but are given to illustrate some of the breadth and depth in these domains. This is the nontrivial content that separates mathematics teachers from people who do math and teachers of other subjects.

1. Teachers of addition and subtraction need to know and recognize Counting Strategies and Additive Reasoning. They also need to know the relationships that lead to the major strategies for addition and subtraction and important models:

   - Important relationships include: the five structure, partners of 10, 10 + single digits = teens, any number plus 10, rounding, multiples of 10, place value patterns, two meanings for subtraction

   - Major addition strategies: Add a Friendly Number, Get to a Friendly Number, Add a Friendly Number Over, Give and Take

   - Major subtraction strategies: Remove a Friendly Number, Remove to a Friendly Number, Remove a Friendly Number Over, Constant Difference

   - Important models: number racks, open number lines, equations

2. Teachers of multiplication and division need to know and recognize Counting Strategies, Additive and Multiplicative Reasoning. They also need to know the relationships that lead to the major strategies for addition, subtraction, multiplication, and division.

   - Important multiplicative relationships include: place value, scaling relationships like times 10, using 5×6 to find 5×60 or 50×6, quotative and partitive division types

   - Major strategies: Smart Partial Products and Quotients, Over/Under, Five Is Half of 10, Doubling/Halving for multiplication, using Quarters and Flexible Factoring for multiplication, Equivalent Ratio for division

   - Important models: open arrays, ratio tables, equations

3. Teachers of Proportional Reasoning need to know and recognize Counting Strategies, Additive, Multiplicative, and Proportional Reasoning. They also need to know the relationships that lead to the major strategies for addition, subtraction, multiplication, division, fractions, solving proportions, and important models.

- Important proportional relationships include: the within and between relationships of fractions, ratios, decimals, percentages with their five meanings: part-whole, operator, quotient, measurement, ratio

- Major strategies: Using Within versus Between, Scaling Up and Down in Tandem to find important equivalencies, Scaling in Tandem to find Unit Rates

- Important models: open double number lines, percent bars, ratio tables, equations

4. Teachers of Functional Reasoning need to know and recognize Counting Strategies, Additive, Multiplicative, Proportional, and Functional Reasoning. They also need to know the relationships that lead to the major strategies for finding rates of change.

- Important functional relationships include: constant and nonconstant (especially exponential) rates of change, dependency relationships, short- and long-run behaviors of parent functions, transformations of functions, periodic behavior

- Major strategies: finding and using differences with polynomial data and ratios with exponential data, shifting proportional relations to find equations of other lines, transforming functions, adding ordinates to determine long-run behavior and limits, using ratios of polynomials to describe rational functions, finding good models of data to interpolate, extrapolate, and describe trends

- Important models: open number lines, coordinate axes, paired number tables, parent functions, tangent lines

For teachers to learn the aforementioned important models and strategies, they need to know the difference between a mathematical model and strategy.

MODELS VERSUS STRATEGIES

I am facilitating a Problem Strings lesson with a group of elementary teachers in New South Wales, Australia. We have solved these problems: 27 – 10, 27 – 9, 43 – 10, 43 – 9, 36 – 20, 36 – 19, 53 – 18. When I ask 27 – 9, a teacher shares that he had used 27 – 10 to help. When I ask 43 – 9, a teacher shares that she had used 43 – 10 to help. Similarly, a teacher shares that she had used 36 – 20 to solve 36 – 19.

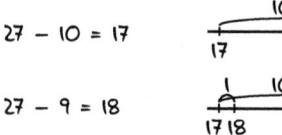

$27 - 10 = 17$

$27 - 9 = 18$

$43 - 10 = 33$

$43 - 9 = 34$

$36 - 20 = 16$

$36 - 19 = 17$

$53 - 20 = 33$

$53 - 18 = 35$

For the last question, 53 – 18, I ask teachers if they could create their own helper problem and they suggest 53 – 20. I then invite teachers to put words to the relationships, to describe the strategies they have been using.

One teacher says, "We've been using a jump strategy. We're jumping on a number line."

She has confused models and strategies. The focus on the representation of the thinking, jumping, rather than the mathematics happening is the giveaway. The fact that she is jumping on a number line is about a model, not a strategy. *How* she chooses to jump on the number line is her *strategy*.

To help turn their attention to the way they are using relationships to jump, which means their strategy, I respond, "But *how* have we been jumping?"

"What do you mean?" she asks.

"How would you describe the pattern you saw in the String that helped you create your own helper problem? When you chose to use 53 – 20, a problem not given, to help you find 53 – 18, why did you decide to use that problem? How does 53 – 20 relate to 53 – 18?"

"Ah, okay, we had been using problems that subtracted too much, that were easy to solve. So I just made a problem that was jumping too far, then adjusted."

Now we're talking strategy! We're talking about how we're using the relationship between the 20 we've chosen and the 18 we're supposed to subtract. Subtracting by a friendly number that's too big and then adjusting—that's a major subtraction strategy. The number line is the model we are using to make that thinking visible.

TRY IT

In Figure 7.2, do you see two different models or two different strategies for solving 18 × 35? How do you know?

FIGURE 7.2 ● Two Different Models or Two Different Strategies?

18 × 35

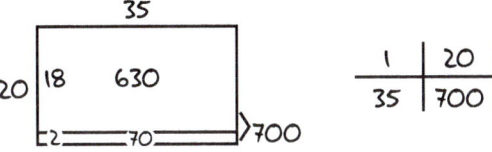

Figure 7.2 shows two different models for using an over strategy to find 18 × 35. And here is a third model, an equation to represent the same strategy:

$$18 \times 35 = (20 - 2) \times 35 = 20 \times 35 - 2 \times 35 = 700 - 70 = 630.$$

It is crucial to distinguish between strategies and models so we can help students focus on their strategies, how they are thinking about the relationships and the connections they're using to solve problems.

MATHEMATICAL MODELS

Making student thinking visible using mathematics models is important. These models can also become tools for reasoning.

Figure 7.3 shows the major models to use in developing mathematical reasoning.

FIGURE 7.3 ● Major Models in Developing Mathematical Reasoning

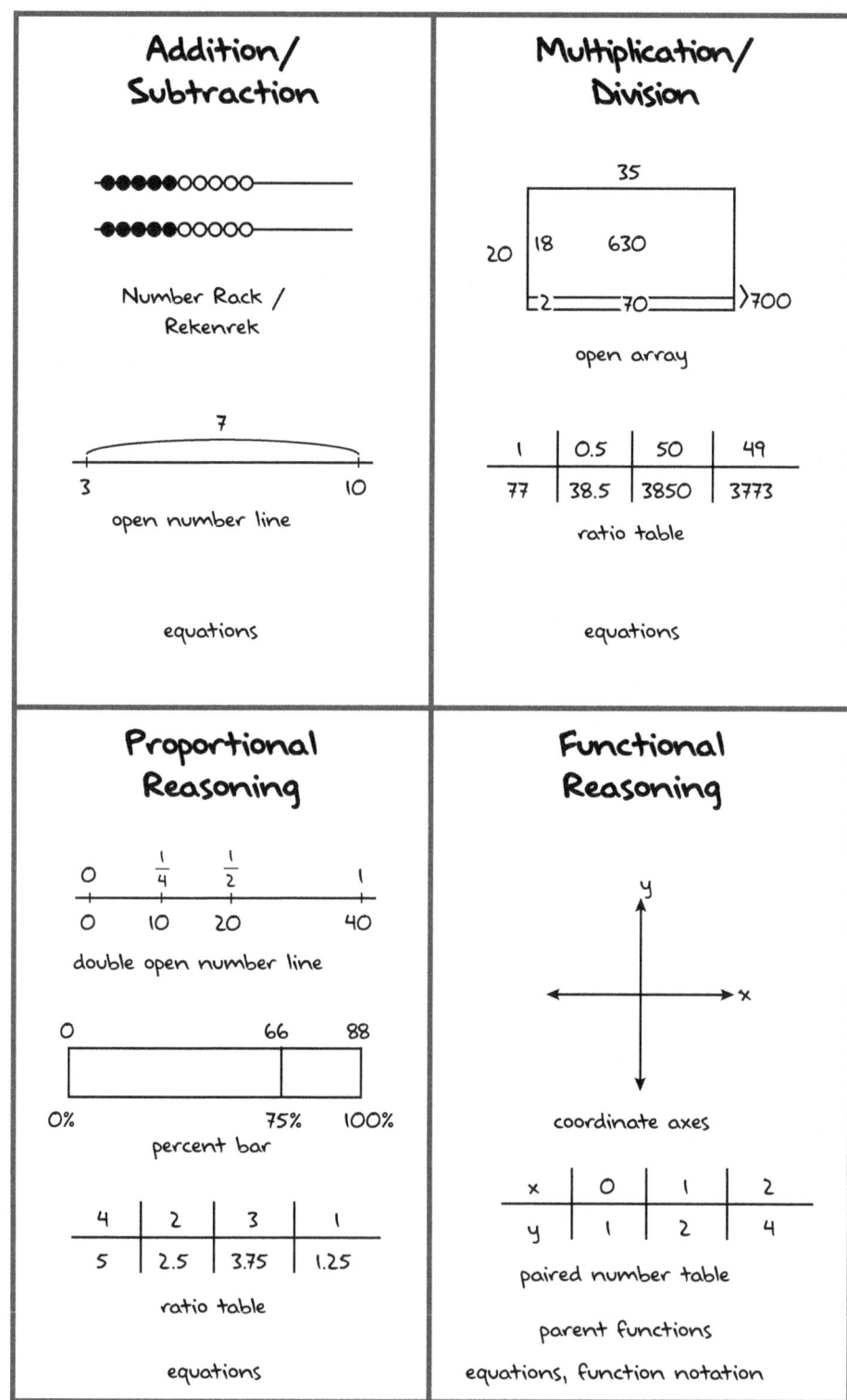

Elementary mathematics researcher Cathy Fosnot says, "Although the emphasis is on the development of mental arithmetic strategies, this does not mean students have to solve the problems *in* their heads—but it is important for them to do the problems *with* their heads!" (Fosnot & Uittenbogaard, 2007). If you're doing the steps of an algorithm in your head, that's not mental math. If you're writing to keep track of your thinking, to help you think, or to represent your thinking, that is mental math. Mental math is not about using your mimicking brain; it's about using your reasoning brain, whether you're writing or not. It can be very helpful to write to keep track of your mental thinking with a model.

Models are important because they enable teachers to elicit and represent student thinking, which provides the avenue for students to begin to represent their own thinking. This self-representation is one of the most powerful tools we have to help students get metacognitive about their own problem-solving process and improve on it.

If models are the way we represent thinking, the language we use to teach and communicate our problem-solving approach, the major strategies are the subject matter we need to communicate.

MAJOR STRATEGIES

There is only a small set of *major* mathematical strategies to develop. Major strategies are defined by the following criteria:

- These strategies not only solve problems, but learning them builds more and more complex reasoning ability in students' brains.

- These are the methods that math-y people actually use to solve problems (Dowker, 1992).

- After examining the set of all problems students need to solve, these are the strategies that will help students be efficient *enough*, giving students access to solve problems that are reasonable without a calculator.

- These are the mathematical relationships students need to progress and succeed in higher math.

Examples of the major strategies are listed in Chapters 3, 4, 5, and 6.

Q: Why didn't my favorite strategy(ies) make your list?

A: I get it, there are many fun/cool ways to play with numbers and relationships. My list consists only of those strategies we need. Without these, we'll leave holes, whether it's in reasoning or too many problems being inefficient to solve. If students want to play with other cool strategies (relationships, not tricks), go for it—but don't overwhelm them. Emphasize the important strategies. If you ensure students leave you having developed the major strategies, they will be prepared for what comes next.

Q: So, now we need to have our students memorize multiple strategies? Pam, they have a hard enough time getting one algorithm down. Now you want students to learn more?

A: It's not about rote-memorizing more things. It's about rote-memorizing less things! Building strategy is about developing brains to reason more complexly, and it is a cyclical process. As we high-dose pattern students, we ask students to solve problems using what they know, we elicit their thinking; we make that thinking visible so it's discussable and comparable. As students verbalize their thinking, see it made visible with a visual model, and describe the relationships, those patterns and relationships become more clear and connected to things they know. With that clarity, they use the connections to solve more or new problems, and the cycle continues. Identifying and describing a strategy is really part of the clarifying process as students build those mathematical relationships.

These strategies are not new or radical. You will find them in many traditional textbooks, but usually in one of two ways.

One, they will be accompanied by the traditional algorithms and almost always with the algorithms receiving more weight, being more important, set up as the goal.

Two, these strategies will be listed in the midst of many other not important or less sophisticated strategies with the impression that any "alterative" strategy is fine, that they all carry the same weight of importance. "Learn" a different strategy for five days in a row. Counting Strategies are right next to Additive Strategies, and then models are thrown into the mix as if they are strategies. It's a mixed-up, mashed-up list of "mental" methods.

Both of these treatments leave many teachers wondering, why bother at all? If the goal is the algorithm, let's just get to it. They know students will need lots of practice to get the algorithm right. Why waste time with these other things that we don't need anyway?

Remember the goal of math class: to build mathematical reasoning. The major strategies get that done.

FREQUENTLY ASKED QUESTIONS

Q: But don't students need a general solution?

A: If they do, they can use a calculator/computer. And if they've built mathematical reasoning, they'll know what to put in that calculator. Because, what works better to develop mathematical reasoning—having students mimic a general solution, or using strategies to develop important mathematical relationships?

There is a pervading cultural perception that students need to have a general solution method, one that can solve even the gnarliest of problems. Every subtraction problem can be solved by borrowing, every fraction division problem can be solved by invert and multiply, every proportion can be solved by cross multiply and divide. No matter the numbers, there is a general algorithm.

The by-product of this general solution perception is that we teach those algorithms. Students need to solve $\frac{19}{52} + \frac{31}{17}$, so let's give them the process to follow. We think it's important to find 64.7% of 17.3, so here's a procedure. In case students ever need to find the rate of change between (4.57, −0.53) and (−14, 432), we'd better drill the steps.

What if solving problems like these with a rote-memorized procedure isn't the best way for students to *learn* fractions or percentages or rates of change? It is not.

The fallacy is not the general solutions themselves, and it's not that creating general solutions isn't useful. Algorithms are amazing human achievements. The fallacy is believing that rote-memorizing and mimicking a general solution equate to proficiency of the domain in which the general solution solves problems.

We don't need to learn to mimic a general solution *procedure*, because we have calculators/computers. We need a general *learning* solution. That learning solution is to high-dose pattern students to develop mathematical relationships that lead to strategies, all of which builds mathematical reasoning.

There is a difference between a formula (a general description that expresses relationships like $y = b + mx$) and a step-by-step procedure. My colleague Kourtney Peters says, "A formula doesn't initiate a to-do list." When we are dealing with a formula, we are seeking to understand the relationships involved, what is happening with the phenomena. For example, how a change in b affects the graph of the line. Similarly, we need students to understand that the "=" symbol doesn't mean "do it." Rather, the "=" symbol expresses the relationship of equivalence (Empson & Levi, 2011).

FREQUENTLY ASKED QUESTIONS

Q: But, Pam, I have to really think hard sometimes when I'm using a strategy. It can be so much easier to just mindlessly repeat the steps of an algorithm.

A: I believe you. First, remember all of the time it took you to get those steps to become automatic. Let's use that time better! It goes back to the purpose of mathematics class. If the purpose is to easily get answers without thinking, then just use a calculator. If the purpose is to develop mathematical reasoning, then thinking and grappling is the price to be paid. Remember, I didn't say that math is easy. I said that math is figure-out-able!

Once we as teachers know the difference between strategies and models, and we can identify and use the major strategies, then we can focus on making that thinking visible, comparable, and discussable. Which leads to our next stage, eliciting and representing student thinking.

ELICITING AND REPRESENTING STUDENT THINKING THROUGH PURPOSEFUL TASKS

Once you're clear on solving problems using what you know and are clear on models versus strategies for the content you teach, you're ready to learn to elicit those strategies from students and represent their thinking to make it visible.

ELICITING AND REPRESENTING STUDENT THINKING

Focusing on and representing student thinking is not trivial for many reasons.

For one, most of us were not in classrooms where our thinking was solicited. We lack the experience to draw upon. In the moment-to-moment decision making, it is natural to lean back on our prior experiences. We need new experiences.

Another reason is that representing student thinking requires the ability to listen carefully to students, to understand what they are saying and how they are using relationships, instead of superimposing our ideas on them. This takes concerted effort and practice.

Also, understanding student thinking demands math content knowledge for teaching. An engineer solving a problem and an architect designing a building do not necessarily have to understand someone else's method in order to get their job done.

Teachers, however, need to own much more than one way to make sense of students' developing as-yet-unsophisticated attempts. And where to nudge next.

Teachers also need to choose an appropriate model to make student thinking visible. Some models are better to develop a beginning *spatial* sense for a strategy, and then a different model is better to use *as a tool* to solve those problems. For example, a ratio table works well for students' scaling strategies in beginning multiplication, and the open array is fantastic for building spatial sense for area and chunking areas for multiplication strategies, but later we want to use a ratio table as a tool to solve complex multiplication, division, and Proportional Reasoning problems. Teachers also must decide how they will represent the thinking on that particular model.

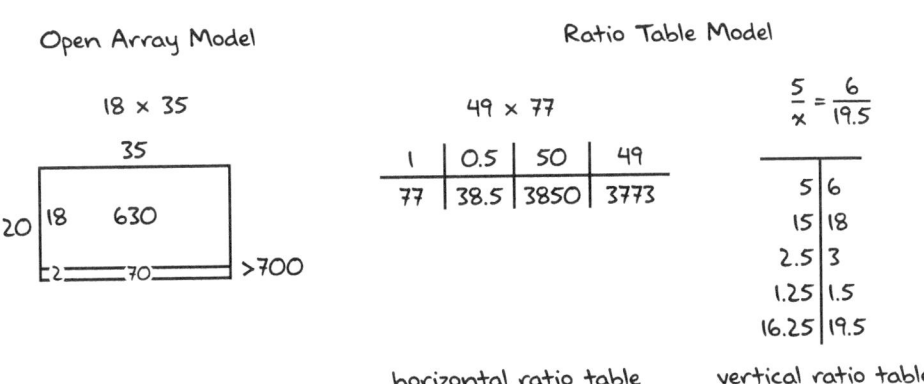

However, not all strategies can be represented with each model. For example, the doubling/halving multiplication strategy is represented well with open arrays and equations, but cannot be represented on a ratio table as there are no equivalent ratios involved in that strategy.

18 × 35 Doubling Halving Strategy

$$\div 2 \left(\begin{array}{c} 18 \times 35 \\ 9 \times 70 = 630 \end{array} \right) \times 2$$

equation model

open array model

When you were reading each of the Problem Strings we walked through in Chapters 2–6, you could see this modeling process:

- Students share their thinking about how they solve a problem.

- As the teacher represents the thinking, students see their and their classmates' thinking made visible. They can clarify their thinking as they compare what their brain is doing with how it looks visibly.

- Students gain more clarity as they discuss, compare, and generalize the patterns that are happening and being used.

- With time and experience, students begin to use certain models as tools for thinking. These important models are number racks, open number lines, open arrays, ratio tables, coordinate axes, and equations.

TIP

The models used in this book to exemplify each strategy were deliberately chosen as good models for that strategy. Use these examples to help you determine which model to use.

FREQUENTLY ASKED QUESTIONS

Q: Is this the same as C-R-A, concrete-representational-abstract?

A: No, it's not. Our modeling process is not about giving students steps to do with manipulatives, then drawing those steps in a picture, and then writing those steps with numerals or other symbols. The modeling process described in this book is more about natural development, where seeing, discussing, and comparing the results of efforts inspire adjustment. Because students are able to see their thinking made visible, they can point at parts of it, make generalizations, and develop stronger mathematical relationships. Our modeling process is often confused with giving students manipulatives to "see" the math. Students do not magically understand the mathematical relationships when they are holding manipulatives. Students do not learn math by mimicking steps using manipulatives. Students learn to mathematize as they solve problems, and teachers help make that thinking visible using manipulatives and models.

Teachers learn to elicit and represent student thinking as they engage students in purposeful mathematical tasks.

PURPOSEFUL TASKS

In Chapters 2–6, I provided some activities to get you started in the Try It in Your Classroom features. These are worthwhile ways to get started: trying things that help build your sense of what it means to math the way mathematicians math. They are a good part of a well-rounded classroom.

But, of course, those tasks are not enough. What does math-ing in a real math classroom look like?

A progression of Problem Strings and Rich Tasks, with other supporting tasks like Problem Talks, games, and other instructional routines, is the most effective way to round out your classroom.

Although many programs advocate Problem Talks or Rich Tasks as the most important or the first task to try, we suggest that teachers start with the instructional routine called Problem Strings.

THE BEST STARTING TASK: PROBLEM STRINGS

One of the best ways to start teaching real math-ing—getting students to solve problems, reason using what they know, and build important relationships toward strategies—is to implement the instructional routine called Problem Strings. Problem Strings provide students with the high-dosage patterning they need to make mathematical connections.

Thank you to Brendan Scribner for the description of "high-dosage patterning."

In each chapter, you've seen examples of students solving a series of related problems that build relationships intentionally toward a strategy, model, or big idea. These series of problems, called Problem Strings, are fantastic tools to develop mathematical reasoning:

- The string of problems is an intentional, purposeful sequence of related problems that has a mathematical goal.
- The teacher gives each problem, one at a time.
- Students are encouraged to solve the problems any way they can reason about them.

TIP
Some models are not good to use as tools for computation as they can keep students using less sophisticated reasoning than the problems call for. Do not suggest that students use these models to compute. Some examples include: ten frames, closed number lines, hundreds charts, base-ten materials, or gridded coordinate axes to find slope.

- As the teacher represents the thinking, students see their and their classmates' thinking made visible. They can clarify their thinking as they compare what their brain was doing with how it looks visibly.

- Students gain more clarity as they discuss, compare, and generalize the patterns that are happening and being used.

- Because there is a series of problems, intentionally sequenced, students consider the same relationships over and over again, traveling that mental path until the strategy becomes a natural outcome.

- Students solve many problems in a short period of time, with instant feedback. They can adjust and try again on the next problems, over and over again, with increasing confidence, accuracy, and efficiency.

- Used in a series of multiple Strings across multiple lessons, Problem Strings help students strengthen their mathematical relationships over time with multiple experiences as the problems get more complex.

Problem Strings are extremely helpful as a common structure that all teachers can use across all grades. Teachers at every grade level can have conversations about how they are choosing the model to use; which strategies they are having students share, in what order, and why; how to decide whom to call on; how and when to anchor the learning, etc. It doesn't matter that the math content varies; all teachers can gain from discussing these choices.

 TIP

Often, context is extremely helpful in a good Rich Task. If done well, students stay in context, reasoning about packs and sticks of gum, slices of pizza and cost, not about contextless numbers or decimals. If you find students leaving the context behind too soon or too much, consider that either the task is not truly problematic or that your facilitation might have trivialized the context and inadvertently encouraged students to abandon the context.

RICH TASKS

The kind of tasks I prefer, I'll call Rich Tasks. Cathy Fosnot and Maarten Dolk, in their *Young Mathematicians at Work* series, describe the best tasks as a "truly problematic situation." A task does not need to be *real* to be worthwhile; rather, the mathematics needs to be *realizable* (Fosnot & Dolk, 2001). Realistic Mathematics Education discusses *realistic* as: imaginable, realizable, not just "real-world" (Freudenthal, 1987; van den Heuvel-Panhuizen & Wijers, 2005), where the situation is "begging to be organized" (Gravemeijer & Terwel, 2000, p. 787) from the students' perspective.

More detail about Rich Tasks is out of the scope of this book, but I will suggest two principles:

1. The purpose of a truly problematic situation is *not* to just get students engaged, then maybe get some conceptual understanding, give them an algorithm

to solve the problem, and then the teacher reveals the "answer" to see if it worked.

2. In a good, well-facilitated Rich Task, actual learning occurs during the solving of a Rich Task and during the following math discussion as teacher and students co-negotiate meaning together.

The best Rich Tasks are those that intrigue students, have multiple solution pathways, and help develop important big ideas of mathematics.

FREQUENTLY ASKED QUESTIONS

Q: Why isn't there more in this book about problem types and word problems?

A: Different problem types are important because they influence how students will begin to solve word problems (Carpenter et al., 2015; Empson & Levi, 2011). It is therefore important to understand the different types, when and how to use them. However, since this book focuses on the strategies alternative to algorithms, the examples are often restricted to "naked" problems so as not to conflate or associate a problem type with a particular strategy. More work around problem types and solving word problems is forthcoming in grade-band-specific books that will be releasing every six months in 2025 and 2026 from Corwin. Stay tuned!

To facilitate Rich Tasks well, teachers need to develop and capitalize on high leverage teacher moves.

DEVELOPING HIGH LEVERAGE TEACHER MOVES

Once teachers can solve problems using what they know, own the major models and strategies for the content they teach, and elicit and represent student thinking during important tasks, they often turn their attention to facilitating those tasks better and better. They have bandwidth to consider the pedagogical moves they make during teaching that encourage mathematizing.

As Cathy Fosnot suggests, "Teachers walk the line between planning for the community and being a part of the community" (Fosnot & Dolk, 2001). This means that teachers have the responsibility to know the content, direct the learning to mathematical goals, and help students work together. But teachers also join the community by listening carefully to students, being intensely interested in students' thoughts and ideas, joining students in their wonderings, and crafting the discussions to create a community of learners who are co-creating knowledge together.

To accomplish these goals, teachers can develop and use certain high leverage teacher moves.

Some of the more familiar high leverage teacher moves include:

- Questioning strategies
- Patterns of questioning: focusing versus funneling (NCTM, 2014)
- Wait time
- Differentiating for all learners

The following top my recommendations for the most impact.

PRIVATE RESPONSE SIGNALS

Requesting discrete, private response signals is preferable to public hand-waving. A thumbs-up held close to the chest is easily visible from across the classroom as you are facilitating a Problem String, but it isn't immediately obvious to other students, which means the learner can communicate their readiness but remain discreet to their peers. Such a private signal allows you to manage the flow and timing by considering the entire class's progress while not stressing students who need more time and may feel pressured by waving hands.

> ### TRY IT
>
> During your next classroom interaction, notice how students are responding to questions. Is it hands up or waving? Calling out? How many students are answering? How much time do students have to think about their response before hearing someone else's? Are you satisfied with the time and space your students have to think? If you think students could benefit from more time, request that students give you a private response signal.

USING A NEUTRAL RESPONSE

If you as an educator keep a neutral response, in facial expressions, body language, and tone of voice, this helps reinforce your classroom's focus on reasoning instead of answer-getting.

If you reserve your smiles or scowls for feedback when students give an answer, you are signaling answers are what you really care about. I'm not suggesting that answers do not matter, but that cueing immediately whether a student is correct or not is extremely effective at shutting down student thinking.

If a student knows they are wrong, they are then much less likely to want to explain their thinking. If they know they are right, they may see no need to justify their idea. If they get in the habit of justifying their thinking, the correct answers take care of themselves and reasoning grows.

> ## TRY IT
>
> During your next classroom interaction, take note of your responses to students. Do you immediately cue students if they are correct or incorrect? Do students wait for your cues? Are students in the habit of justifying their answers? Try responding more neutrally, with curiosity about why they are thinking that way. Ask students how they know.

JUST IN TIME VOCABULARY VERSUS JUST IN CASE

Tagging vocabulary terms when students have a need to use the language, in the midst of the learning, is far more effective than front-loading vocabulary when students still lack the context to understand and make use of it.

For example, the best time to define *slope* is much closer to that moment when they are trying to verbally describe what they are observing when walking toward or away from a motion sensor. In the midst of the experience, the word *slope* is more meaningful than recording it in a notebook ahead of time. Such an approach makes vocabulary useful instead of burdensome.

> ## TRY IT
>
> Think about your next unit and the vocabulary that is helpful in that unit. Plan opportunities that beg the use of that vocabulary. Allow students to use less precise mathematical vocabulary as they describe what they are trying to say. As they do, use the precise mathematical vocabulary. Say, "Ah, we call that a ratio/slope/scaling." Or, "You mean when the line is horizontal/the opposite sides are parallel/the root is the x-intercept."

CELEBRATING REASONING

Everything in your classroom, from how you handle home-work to grading tests to seating arrangements, should reflect that you value reasoning first and not merely answer-getting. You're probably already doing some of this—for example, not giving credit if a student obtained an answer by using a calculator when they weren't supposed to. We simply need to extend that same logic. An answer obtained without learning is not an answer worth getting, not in a classroom.

Rooting out all your old practices that focus solely on answer-getting can be trickier than it seems. Remember that most likely the vast majority of your classroom experience, all the systems you grew up with, were designed to value answer-getting and mimicking.

If a student reverts to a memorized series of steps, ask them to listen to classmates' reasoning. Be sure to celebrate that reasoning by discussing it and comparing it to other strategies. Invite students to convince the class of their strategy. Make the conversations about strategy and reasoning. Students will rise to the occasion when they realize your classroom is about mathematical reasoning.

Building Powerful
Mathematics
Workshops

qrs.ly/tvgdthb

As the Math Is Figure-Out-Able team has created online asynchronous Building Powerful Mathematics workshops, we have found it highly effective to structure each workshop to first build teachers' mathematics and their mathematics for teaching. Then, the workshop presents montages of short video segments, taking teachers back through important moments of the workshop to examine through two different lenses: one through the lens of tasks (purpose, strengths, when and how to use) and the other through the lens of developing high leverage teacher moves. Check out a Building Powerful Mathematics *workshop here: https://www.mathisfigureoutable.com/workshops.*

These important teacher moves support students in developing as mathematical reasoners and help the teacher move the math forward.

Once teachers reach this stage, they turn their attention to sequencing tasks and advancing the math by creating their own tasks.

SEQUENCING TASKS AND ADVANCING THE MATH

Once teachers are solving problems using what they know, owning the major strategies and models for their content, eliciting and representing student thinking, and refining their high leverage teacher moves to facilitate important learning tasks, teachers get really interested in and have the bandwidth for sequencing tasks to advance the mathematics.

Some of those milestones are:

- Write your own Problem Strings, compare with a colleague, tweak as you see results

- Create sequences of Problem Strings

- Create sequences of lessons that include Problem Strings, Rich Tasks, appropriate games, and instructional routines

- Plan the sequence of models to develop across a topic

- Work to assess reasoning, efficiency, and sophistication that is helpful for students to monitor their own learning

- Use properties to explain and compare strategies

Once teachers are creating sequences of lessons to move the math forward for all students in a particular area of mathematics, they choose another area and cycle through the stages.

Mathematics has many areas to explore. As teachers learn more content and continue to learn about their students using their teacher moves we have a perfect storm of mathematizing.

TIP

A great way to get started writing your own Problem Strings is to first facilitate expert pre-written Strings. Get used to the flow and structures. Build your own mathematics for teaching. Then, write "echo" Problem Strings based on those exemplar Strings following their structure and organization but using similar numbers to achieve the same goal. This forces you to analyze the Strings and learn the structures and relationships involved in that particular String. An additional task is to write "next step" Strings that up the ante. These Strings take students from where they ended on the previous String and further the mathematics just enough to maintain access and challenge for everyone. This is wonderful work on which to collaborate with colleagues.

FREQUENTLY ASKED QUESTIONS

Q: What is the place for algorithms in K–12 mathematics education?

A: They don't have an important place in K–12 mathematics education. Too many students get trapped and "find it difficult to change habits once they have been taught the use of algorithms" (Hurst & Huntley, 2018, p. 66; Hartnett, 2015). Very experienced math-is-figure-out-able teachers, ones who know how to avoid all the traps of the algorithms, could theoretically get some benefit from having students explore why algorithms work. Even then, there are many better ways we could spend that time.

Algorithms have a place in computer science coding classes where the objective is to dissect existing algorithms to learn how they work, use them in efficient ways, and craft new algorithms. Students who have built mathematical reasoning will be much better equipped to create these algorithms.

Q: What can we do if our standards call for an algorithm?

A: Many content standards have a student expectation similar to: "The student will use algorithms and strategies to solve [multidigit addition, subtraction, multiplication, division, fraction] problems." If you are in a grade where you have a similar standard, first remember that the purpose of math class is to build mathematical reasoners, which includes fluidly and flexibly solving problems. Work all year long to help students develop the mathematical relationships that lead to fluently using the major addition strategies. As you do, students will be discussing their thinking, comparing strategies for efficiency and cleverness, and generalizing why strategies make sense and when to use them.

At the very end of your year, give students one more method to consider, *not* as a thing to do, but as a study in why it works. Tell students that, just as they have been doing all year by analyzing each other's thinking to understand, they will analyze a method that historically people might have learned. Give them the very first step of the algorithm and ask them to discuss how that might work. They'll think it's odd because it works against their intuition to start with the most inconsequential part of the number, and they will wonder why anyone would do that. Keep that intrigue alive by giving them the next step and keep discussing. Because your students' sense of place value and properties will be so good by this time, they will be able to note how the place value is cleverly hidden behind the scenes, how the steps take advantage of the digits while successfully ignoring the actual magnitudes of the numbers, and how one set of steps could be used to solve any problem—ingenious!

Your goal is for them to successfully analyze the steps and understand what is happening and then say with confidence, "Wow. That is cool, how all of the place value is hidden, and the steps could solve any problem no matter the numbers. So interesting. And complicated. So many steps every time. Why would anyone memorize all those steps, if we can just think about the problem? Why go to all the trouble to do each of those steps every time, if we can just use what we know to solve it?" They will be math-ing so well that they won't be tempted to rote-memorize those steps. They'll own so many relationships that they will confidently use them to reason through the problems. And if they really don't want to reason through a particularly arduous problem, they'll wisely reach for a strategy that will give them an accurate answer—a calculator. They'll know when they have power over a problem and when it makes sense just to reach for technology.

When you analyze an algorithm with students, use numbers that make it easy for students to feel the arduousness and futility in using So. Many. Steps. Here are a few sample problems that have worked well for me.

Addition	Subtraction	Multiplication	Division	Solving proportions	Writing the equation of a line
$37 + 99$	$203 - 98$	25×18	$1188 \div 12$	$3{:}8 = x{:}41.6$	$(0, 4)\ (1, 5)$

Q: But since it's in my standards, won't my students be tested on it?

A: Not from what I've seen, and I've been asking people to send me examples for decades. I have yet to see a released item from a high stakes test that assesses the steps of an algorithm. (Items like this would be way too easy for most students—they would not provide the range of results that test item writers seek when writing a norm referenced test.) I have seen items from test prep publishers where they guess what might be on those tests, but not an actual released item. Now, it's certainly possible that I just have missed one or two. However, even if there are one or two items on a high-stakes test that assess the steps of an algorithm, I would much rather have the high scores on those tests because students are confidently reasoning through the test rather than get meager success for that one item. And since students have been studying different strategies all year long, chances are that they will be able to figure out how to answer such questions, too.

(Continued)

Q: What will happen to my students in their next class if I don't teach them algorithms? Won't that put them at a disadvantage next year?

A: Make sure you send them off with the knowledge that no matter what their teachers tell them next year, math is and always will be figure-out-able. And then consider—the year you gave them of knowing they could math, reasoning using what they know, feeling confident to dive into difficult ideas knowing they could make sense of them—there's a chance that they'll never have a teacher again who will help them develop as mathematicians, but they had a super good year mathematizing with you. Better to have had that single year confidently math-ing than none at all.

Q: What if the teacher next year requires students to use algorithms?

A: If teachers are requiring students to use algorithms, most of these teachers do not actually care *how* students do previous years' math so long as they *can do it*. The third-grade teacher doesn't care how students are adding single digit numbers as long as they do it well. The algebra teacher doesn't ask how students are subtracting or multiplying whole numbers. If students are operating with fractions, the calculus teacher doesn't care how.

There are three isolated groups of teachers who care how students did last year's math because their chosen algorithm uses a previous algorithm.

- Teachers who force the long division algorithm wish students were already good at the subtraction algorithm.

- Teachers who force algorithms for operations with decimals wish students were already good at the operations with whole numbers so they can just add a bit of lining up and/or butt-cheeking (see Chapter 4).

- Teachers who force the long division algorithm for polynomial division wish that students remembered long division with numbers so they could do the same steps with polynomials.

But consider what happens if your students have a super year of building reasoning with you and then get a teacher who demands memorizing and mimicking.

- If the student is a good memorizer-mimicker, they can memorize and mimic next year, with a whole lot more understanding to be able to know *when* to use that algorithm, like word problems.

- If that student is not a good memorizer-mimicker, they will not do well trying to memorize and mimic *just* like they wouldn't have in your class, but they will have made it 1 more year before their grades and confidence tanked.

There is a chance that students who spent the year with you building their mathematical reasoning may feel in the next year that they have to rote-memorize and mimic if they want to get grades from a teacher who demands algorithms. But that cannot take away the good, positive, confident year they had with you, nor can it take away the mathematical relationships they built. They will probably know better than ever which operations problems call for, their place value and reasonableness will be improved, their problem solving will be solid. Rather than 2 years of memorizing and mimicking, better to have had one really good year. *You* have control of this one good year.

Q: What about students with disabilities?

A: I am not a special educator. I do not claim to have that expertise. That said, I have worked with hundreds of adult education teachers, inclusion teachers, and intervention teachers who are the experts and they report time after time that *this* is where their students thrive. In Rachel Lambert's book *Rethinking Disability and Mathematics*, she states, "A key idea of this book is that what students with disabilities need to succeed in math is not qualitatively different than what all kids need." And that "it matters how we think about the goal of mathematics itself. Mathematics is not just memorizing. We should not make kids wait to have fun in mathematics until they have endured school mathematics" (Lambert, 2024, p. 13). I state emphatically that I believe all students can learn and do more real mathematics than they can fake math. And more importantly, learning and developing real math-ing will be more important to all students than mimicking algorithms can ever be.

Q: How much do I need to know before I stop teaching algorithms and start developing strategies?

A: Less than you think! I encourage you to dive in! Start with the Try It activities in each chapter. Join in online talks like #MathStratChat to solve problems and learn how others are solving problems. Ask everyone you meet how they think about problems. Learn to listen to their thinking. Attempt to make their thinking (strategies) visible using a model. Work with colleagues. Try Problem Strings and learn right along with your students.

Conclusion

Remember Jordan? That wonderful thinker who uses what he knows to solve problems? A year later, I do some more videoing with fifth-grade students, asking them some division questions.

As soon as Jordan sits down, I can see that confident, relaxed, reasoning Jordan is replaced with nervous, uptight, retrieving-from-rote-memory Jordan. As I read the word problem to him, he writes down two arrows, one pointing up and one down. He fusses. I wait. He is visibly stressed. I wait. I finally ask, "Jordan, what are you thinking about?"

He looks up at me and says something like, "Big number and small number means...to divide?" I have never heard that rule, but I can imagine it.

I say, "Jordan, can you just think about the problem?"

He sighs heavily and replies, "Oh, Ms. Harris, my teacher told me that I had to memorize this. Is it okay to just think?"

"Yes, yes it is," I say softly as he is already diving in, reasoning about the problem. Now, as he reasons through the problems, he uses fourth-grade strategies. It is obvious he has not been encouraged and enabled to develop any more sophisticated thinking since fourth grade. What is happening in his fifth-grade math class?

Jordan's fifth-grade teacher has been to my trainings. And she has told me on more than one occasion that she doesn't buy what I am selling. She firmly believes that students need to be drilled in the algorithms. To not drill them would be a disservice to her students. I am sure she believes she is doing the best she can for her students. So she continues to drill—and marks answers wrong if they are not found with her algorithms.

Kim has become the campus coach that year. "Kim," I ask, "can you work with students? Jordan seems defeated."

I contact Jordan's mother and ask for permission to work with him outside of class. She responds, "Pam, I guess you can. But . . . we're just not math people. We've been surprised that Jordan did so well for so long. We're not surprised that he's failing fifth-grade math."

For more on avoiding non-growth-mindset language, see Carol Dweck's Growth Mindset *(2006), Jo Boaler's* Mathematical Mindset *(2022), and Vanessa Vakharia's* Math Therapy™ *(2024).*

With her permission we move forward. Kim is able to make some space in her day to pull Jordan and continue mentoring him to mathematize. Jordan is happy to stop trying to mimic. At the end of that year, he fails fifth-grade math (still being graded on mimicking) but scores super high on our state high-stakes test—using reasoning.

Checking on him in the next year, I ask his sixth-grade math teacher if I can video Jordan. He sings Jordan's praises. "Great kid. He is a thinker. Love having him in class. He's doing fantastic."

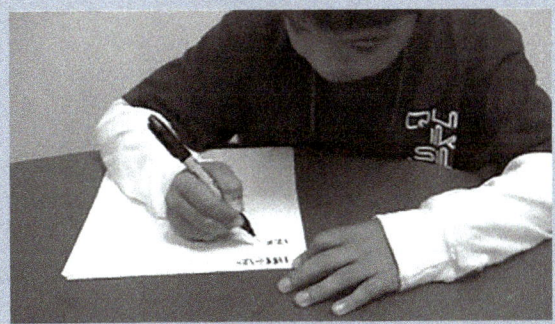

We sit down in a nearby office. "Jordan, what is 1188 divided by 12?" Before you read further, think about that problem and predict what strategy he might use.

Jordan writes the problem and then "12 × " and pauses. Then he finishes "12 × 100 = 12,00 [*sic*]."

Of course, the comma placement is not quite correct, but I'll bet he is thinking in place value terms, 12 × 100 is 12 hundred, and so it makes sense to him to put the comma after the 12. We can work with that.

He pauses and then continues: "12 × 100 = 12,00 − 12 = 11,98, [*sic*]."

He sits up with a big smile. "I did it. It's 99."

Yes, yes he did.

That whole process takes exactly 1 minute. Sixty seconds of reasoning. To solve 1188 ÷ 12, Jordan doesn't reach for a step-by-step procedure. He asks himself what he knows about 12s and he knows 12 × 100 is 1200. But he doesn't need 1200, only 1188. How far is 1188 from 1200? Just one 12! Bam, he only needs 99. Jordan is using an over division strategy, finding a friendly dividend that is a bit too big and adjusting. This is Multiplicative Reasoning with division.

To watch Jordan in action with division, see the QR code on this page.

A few years later, I ask Jordan if we can record him again. Before we turn the camera on, I ask his mother how things are going in his math classes. She replies, "Oh, he could be doing better if he put more effort into it."

"Mom," he sheepishly inserts, "I'm super busy. And I am getting a B."

"Yeah, well, he could be getting an A if he put in more time." She smiles as she walks out of the room.

Watch 6th-grade Jordan math-ing with division

https://qrs.ly/ 7mg3rj4

Remember that this is the mom who gave us permission to work with fifth-grade Jordan, but didn't hold much hope, because they were "just not math people." Notice the change! As Jordan has continued to figure out math, she is realizing he *can* figure out math. The low expectations have been replaced with hopeful, high expectations.

I first asked Jordan if he'd ever thought about fractions using a clock model. He looked a little concerned and said, "It's been a long time since I've done fractions."

No problem. We have a short one-minute back and forth about $\frac{1}{2}$ on a clock: 30 out of 60 minutes, two 15-minute chunks out of four 15-minute chunks, six 5-minute chunks out of twelve 5-minute chunks.

I ask, "So Jordan, using that clock model, what would $\frac{1}{4}$ plus $\frac{1}{3}$ be, $\frac{1}{4}$ plus $\frac{1}{3}$?"

Think about that $\frac{1}{4} + \frac{1}{3}$. What might he do, thinking about a clock?

Before I can finish the question, he says, "That'd be seven twelfths."

"How'd you do that?" I ask.

He replies with a smile, "Because $\frac{1}{4}$ is, uh, 3 o'clock and $\frac{1}{3}$ is 4 and 3 + 4 is 7, so $\frac{7}{12}$." He uses the relationships on a clock, that $\frac{1}{4}$ of an hour is three 5-minute chunks and $\frac{1}{3}$ of an hour is four 5-minute chunks, so adding those together means seven 5-minute chunks, which is seven out of the twelve 5-minute chunks.

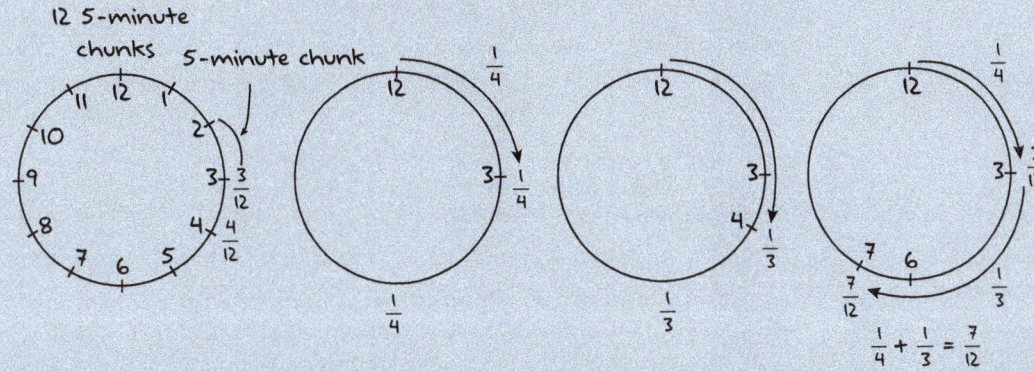

To watch Jordan in action reasoning about fractions, see the QR code on this page.

Two years later, Jordan's mother calls me. "Pam, Jordan's been doing great, but now he wants to take calculus. Calculus! What do you think?"

You go, Jordan!

Watch 10th-grade Jordan math-ing with fractions

https://qrs.ly/ 6hg3rj5

We learned an important lesson from Jordan. When he left Kim's fourth-grade real mathematizing class to go into a rote-memorizing mimicking class, we assumed students would take with them the message: No matter how their next teacher taught it, math was still figure-out-able! Once we realized the importance of declaring that message and got Jordan on a figure-out-able course, he didn't buy into the fake math myth. He continued to figure it out. That's not to say that it was easy. It would have been far better if he'd had teachers who actively helped him figure things out. But not trying to rote-memorize and mimic allowed him not only to go farther in math but also to maintain his identity as a mathematician.

Recently, I ran an online challenge, inviting teachers all over the world to learn more about teaching real math-ing. Jordan was one of my guests.

In his words:

I had a teacher [Kim] that really cared about the way I was thinking. I remember she would come and sit down with me and look at me, and be like, we're figuring this out. Show me how you did this. I want to see how you did it. And she made sure that I got it, but she made sure that not only did I get it, but that I was confident in the way I got it. And I was able to show my work and explain my reasoning. And throughout school, even into college, she's always been my favorite teacher. Because of the lasting impact, I always think back to her. Kim Montague really did make a difference in my life. I can do things. I can figure out hard problems. And just because something's new doesn't mean it's bad. But you know, we can still figure it out. It is figure-out-able!

To hear from adult Jordan, watch the video at the QR code on this page.

Watch adult Jordan talk about the power of developing mathematical reasoning

https://qrs.ly/ycg3rj7

My third son, Craig, had it the best in elementary school. He had those elementary teachers after we'd been working together for a while. We were pretty good by that point. He was thinking, reasoning, making sense of all the things. He was math-ing well. And he was able to keep math-ing despite his middle school and high school teachers teaching traditional algorithms. He came home one day in high school. "You know, I get what's going on with those steps my teacher keeps trying to get us to use. I can understand what's behind them and why it works. It's actually really amazing—there's so much going on behind the scenes, things that you can't easily tell while you're doing the steps. It's quite ingenious. But . . . why would anyone memorize and do all of those steps, when you can just think about it?"

Why, indeed?

Later, in the middle of a computer science class in his last semester at Brigham Young University called "Algorithms," Craig called me. "Mom, you're so right!" He and I had been talking about his experience as a computer science major, and this class in particular. "About half of the students in this class are like you were, they memorized and mimicked algorithms. The other half are like me, we reason through stuff. Mom, because we actually understand and use relationships, we can *create* algorithms. Those other students, not so much. They are really struggling to write their own algorithms because they have only ever repeated someone else's."

All students can do more real math than fake math. Math is figure-out-able, and we can teach it that way!

Discussion Questions

1. What's the difference between a strategy and a model? Why does this difference matter?

2. What's the difference between a strategy and an algorithm? How does developing strategies differ from teaching an algorithm?

3. What are the major strategies and models for the content you teach?

4. What is a Rich Task to you? What is your favorite Rich Task? Why?

5. What do you know about Problem Strings? What more do you want to learn about them?

6. Where do you find yourself: working to solve problems using what you know, learning the major strategies and models for content you teach, learning to elicit and represent student thinking, developing your teacher moves to use Problem Strings and Rich Tasks, or sequencing tasks to move the mathematics forward?

7. What is one next step for you in your journey?

TRY IT IN YOUR CLASSROOM

#MathStratChat: An Example of Problem Talks and Problem Strings

Several years ago, I began posting accessible, interesting problems once a week on social media for people to solve and chat about their mathematical strategies. I called it #MathStratChat, and the purpose was to showcase to anyone who followed the conversations that math is figure-out-able. This is like a Problem Talk (Number Talk) where people compare strategies for a problem. It has since morphed to where I now often post series of problems, one a week for a few weeks, so that now the problems have a higher chance of helping people notice the relationships and patterns and build strategies. It's like a slow reveal Problem String!

You can use #MathStratChat as a repository of problems, learn and practice mathematical strategies, and interact with other problem solvers around the world. You can also access prior #MathStratChat

(Continued)

Math Strat Chat

qrs.ly/y1gdthg

(Continued)

problems and view the Twitter/X conversations for each at https://www.mathisfigureoutable.com/mathstratchat.

Purpose

Build your own and your students' strength and flexibility in math-ing.

Routine

Engage in solving problems!

- Create or choose a problem worth solving that is accessible *enough*. To join in as a participant, search #MathStratChat on your favorite social media platform. As of this writing, that includes Facebook, X, and Instagram.

- Present/consider the problem. Give people enough time and space/take enough time and space to grapple with the relationships involved.

- Read or listen to and consider others' strategies. Compare them to your own.

- Share your strategy. Try to represent it in a visual way. Discuss your thinking.

- If a strategy strikes you as desirable to own, send the message to yourself, "I have access to that. I want my brain to do that next time."

- Repeat, ideally with more problems someone has deliberately chosen because the sequence lends itself to developing important mathematical relationships.

Important to Consider

As great and fun as #MathStratChat can be, it's not as powerful in building mathematical strategies as are Problem Strings. That's why I began a couple of years ago to do sequences of problems in social media, related problems from one week to the next. People began to report, "Ah, now I can make that work. I saw the strategy the last couple of weeks and this week it occurred to me!"

Using the high-dosage patterning of Problem Strings is more effective than just one problem at a time, so use Strings of problems more often than just one problem. One problem can be used to assess how you or your students are doing with particular strategies.

Extensions

- Try to find other problems that would use the same strategy you're grappling with. For example, if the problem you solved was $36 + 27$ and you are playing with factoring to find an equivalent problem, like $36 + 45 = 4 \times 9 + 5 \times 9 = 9 \times 9 = 81$, you could seek for other addition problems that have common factors in the addends, like $64 + 24$. How could you use 8s to solve that problem?

- Find problems for which your strategy would be a terrible strategy. For example, using an over strategy to add $43 + 31$, or doubling/halving to multiply 13×7, or using quotative division to find $\frac{4}{5} \div \frac{16}{17}$.

- Find problems that are rich, where there are multiple nifty strategies to solve, like $28 + 49$, 15×18, or finding the equation of the line between $(1, 3)$ and $(2, 5)$.

- Find problems that are pointed toward a specific strategy, where that target strategy is just so good, like anything plus 9, anything minus 9, $99 + 47$, 99×47, $\frac{1}{2} \times \frac{4}{5}$, using quotative division to find $3\frac{1}{5} \div \frac{1}{5}$ or finding the equation of the line between $(-2, 2)$ and $(56, -56)$.

References

PREFACE

Ball, D. L., Thames, M. H., & Phelps, G. (2008). Content knowledge for teaching: What makes it special? *Journal of Teacher Education, 59*(5), 389–407.

CMP (the Connected Math Project). (2018). Retrieved August 10, 2024, from the Connected Mathematics Project, https://connectedmath.msu.edu/

COMAP. Consortium for Mathematics and Its Applications. Retrieved August 10, 2024, from https://www.comap.com/

Common Core State Standards. (2010). *Mathematics standards.* Retrieved August 10, 2024, from https://www.thecorestandards.org/Math/

Connally, E., Hughes-Hallett, D., & Gleason, A. M. (2000). *Functions modeling change: A preparation for calculus.* John Wiley & Sons.

Crayton, D. (2026). *Shining a light on reading, writing, and mathing: Leveraging literacy for mathematical understanding.* Corwin.

Education Development Center, Inc. (2001). *Math in context.* Encyclopaedia Britannica.

Everyday Mathematics. Retrieved August 10, 2024, from https://everydaymath.uchicago.edu/

Fosnot, C. T., & Dolk, M. (2001). *Young mathematicians at work: Constructing number sense, addition, & subtraction.* Heinemann Educational Books.

Harris, P., Murdoch, J., Kamische, E., & Kamische, E. (2017). *Discovering advanced algebra.* Kendall Hunt.

Kamii, C., & Dominick, A. (1998). The harmful effects of algorithms in grades 1–4. In L. J. Morrow & M. J. Kenney (Eds.), *The teaching and learning of algorithms in school mathematics* (pp. 130–140). NCTM.

Kazemi, E., & Hintz, A. (2014). *Intentional talk.* Stenhouse Publishers.

Klein, A. S., Beishuizen, M., & Treffers, A. (1998). The empty number line in Dutch second grades: Realistic versus gradual program design. *Journal for Research in Mathematics Education* (29)4: (July).

Lambert, R. (2024). *Rethinking disability and mathematics: A UDL Math classroom guide for grades K–8.* Corwin.

Liljedahl, P. (2021). *Building thinking classrooms.* Corwin.

Ma, L. (2020). Knowing and teaching elementary mathematics: Teachers' understanding of fundamental mathematics in China and the United States. Routledge.

NCTM. (1991). *Professional standards for teaching mathematics.* The Council.

NCTM. (1998). *The teaching and learning of algorithms in school mathematics.* Author.

NCTM. (1999). *Developing mathematical reasoning in grades K–12.* Author.

NCTM. (2000). *Principles and standards for school mathematics.* Author.

NCTM. (2006). *Curriculum focal points.* Author.

Smith, M. S., & Stein, M. K. (2011). *Five practices for orchestrating productive mathematics discussions.* NCTM.

CHAPTER 1

Carpenter, T. P., Franke, M. L., Jacobs, V. R., Fennema, E., & Empson, S. B. (1998). A longitudinal study of invention and understanding in children's multidigit addition and subtraction. *Journal for Research in Mathematics Education, 29*(1), 3–20.

Crayton (2026). *Born a mather: Disrupting the notion that mathematics is optional.* Corwin.

Dowker, A. (1992). Computational estimation strategies of professional mathematicians. *Journal for Research in Mathematics Education, 23*(1), 44–55.

Fosnot, C. T., & Dolk, M. (2001a). *Young mathematicians at work: Constructing number sense, addition, & subtraction.* Heinemann Educational Books.

Fosnot, C. T., & Dolk, M. (2001b). *Young mathematicians at work: Constructing multiplication & division.* Heinemann Educational Books.

Fosnot, C. T., & Dolk, M. (2002). *Young mathematicians at work: Constructing fractions, decimals, & percents.* Heinemann Educational Books.

Fosnot, C. T., & Jacob, B. (2010). *Young mathematicians at work: Constructing algebra.* Heinemann Educational Books.

Freudenthal, H. (1968). Why to teach mathematics so as to be useful. *Educational Studies in Mathematics, 1,* 3–8.

Freudenthal, H. (1973). *Mathematics as an educational task.* Reidel.

Harris, P. W. (2011). *Building powerful numeracy for middle and high school students.* Heinemann.

Harris, P. W. (2014). *Lessons & activities for building powerful numeracy.* Heinemann.

Hurst, C., & Huntley, R. (2018). Algorithms and multiplicative thinking: Are children 'prisoners of process'? *International Journal for Mathematics Teaching and Learning, 19*(1), 47–68.

Jensen, E., & McConchie, L. (2020). *Brain-based learning: Teaching the way students really learn* (p. 86). Corwin.

Kamii, C., & Dominick, A. (1998). The harmful effects of algorithms in grades 1–4. In L. J. Morrow & M. J. Kenney (Eds.), *The teaching and learning of algorithms in school mathematics* (pp. 130–140). NCTM.

Liljedahl, P. (2021). *Building thinking classrooms.* Corwin.

Montague, K. (Host). (2021, November 16). Know your content, know your kids (No. 74) [Audio podcast episode]. Math Is Figure-Out-Able.

Pearce, K., & Orr, J. (2020, June 22). Mathematics for human flourishing—An interview with Francis Su (82) [Audio podcast episode]. Making Math Moments That Matter.

Piaget, J. (1974). *Biology and knowledge.* University of Chicago Press.

Plunkett, S. (1979). Decomposition and all that rot. *Mathematics in School, 8,* 2–5.

SERP. (2014). Phil Daro - SERP: Mathematics Common Core Standards. SERP. https://serpmedia.org/daro-talks/

Vygotsky, L. S. (1978). *Mind in society: The development of higher psychological processes.* Harvard University Press.

wolframalpha.com. (2024). *Algorithm.* Retrieved June 22, 2024, from https://www.wolframalpha.com/input?i=algorithm

Wright, R. J., Martland, J., Stafford, A. K., & Stanger, G. (2006). *Teaching number: Advancing children's skills and strategies* (2nd ed.). Sage.

CHAPTER 2

Carpenter, T. P., Fennema, E., Franke, M. L., Levi, L., & Empson, S. B. (2014). *Children's mathematics: Cognitively guided instruction* (2nd ed.). Heinemann.

Fosnot, C. T., & Dolk, M. (2001a). *Young mathematicians at work: Constructing number sense, addition, & subtraction.* Heinemann Educational Books.

Fosnot, C. T., & Dolk, M. (2001b). *Young mathematicians at work: Constructing multiplication & division.* Heinemann Educational Books.

Fosnot, C. T., & Uittenbogaard, W. (2007). *Minilessons for extending addition and subtraction.* Heinemann.

Lamon, S. J. (2020). *Teaching fractions and ratios for understanding: Essential content knowledge and instructional strategies for teachers* (4th ed.). Routledge.

CHAPTER 3

Carpenter, T. P., Fennema, E., Franke, M. L., Levi, L., & Empson, S. B. (2014). *Children's mathematics: Cognitively guided instruction* (2nd ed.). Heinemann.

Karp, K. S., Bush, S. B., & Dougherty, B. J. (2014). 13 rules that expire. *Teaching Children Mathematics, 21*(1), 18–25.

Klein, A. S., Beishuizen, M., and Treffers, A. (1998). The empty number line in Dutch second grades. *Journal for Research in Mathematics Education, 29,* 443–464. NCTM. In Sowder, J., & Schappelle, B. (Eds.). (2002). *Lessons learned from research.* NCTM.

CHAPTER 4

Carpenter, T. P., Fennema, E., Franke, M. L., Levi, L., & Empson, S. B. (2014). *Children's mathematics: Cognitively guided instruction* (2nd ed.). Heinemann.

Fosnot, C. T., & Dolk, M. (2001). *Young mathematicians at work: Constructing multiplication & division.* Heinemann Educational Books.

Harel, G., & Confrey, J. (1994). *The development of multiplicative reasoning in the learning of mathematics* (p. 34). State University of New York Press.

Harris, P. W. (2022). *The most important numeracy strategies.* Math Is Figure-Out-Able. www.mathisfigureoutable.com/big

IES. (2024). *Development of mathematical reasoning.* Retrieved April 3, 2024, from https://ies.ed.gov/ncee/edlabs/infographics/pdf/REL_SE_Development_of_Mathematical_Reasoning.pdf

Karp, K. S., Bush, S. B., Dougherty, B. J., Berry, R. Q., & Larson, M. (2021). *The Math Pact: Achieving instructional coherence within and across grades.* Corwin.

van Galen, F., & Fosnot, C. T. (2007). *Groceries, stamps, and measuring strips: Early multiplication.* Firsthand/Heinemann.

Wolfram Mathworld. (2024). *Long division symbol.* Retrieved April 6, 2024, from https://mathworld.wolfram.com/LongDivisionSymbol.html

CHAPTER 5

Fosnot, C. T., & Dolk, M. (2002). YMAW *constructing fractions, decimals, and percents* (p. 58). Heinemann Educational Books.

Freudenthal, H. (1973). *Mathematics as an educational task.* Reidel.

Hackenberg, A. J., Norton, A., & Wright, R. J. (2016). *Developing fractions knowledge.* Sage.

Lamon, S. J. (1994). Ratio and proportion. In G. Harel & J. Confrey (Eds.), *The development of multiplicative reasoning in the learning of mathematics* (p. 93). State University of New York Press.

Lamon, S. J. (2020). *Teaching fractions and ratios for understanding* (4th ed., p. 3). Routledge.

Tierney, C., & Russell, S. J. (2001). *Ten-minute math.* TERC.

wolframalpha.com. (2024). *Cancel.* Retrieved March 20, 2024, from https://www.wolframalpha.com/input?i=cancel

CHAPTER 6

Connally, E., Hughes-Hallett, D., & Gleason, A. M. (2000). *Functions modeling change: A preparation for calculus.* John Wiley & Sons.

Demana, F., & Waits, B. (1990). *PreCalculus mathematics: A graphic approach.* Addison-Wesley.

IES. (2021). *Development of mathematical reasoning.* Retrieved August 3, 2024, from https://ies.ed.gov/ncee/edlabs/infographics/pdf/REL_SE_Development_of_Mathematical_Reasoning.pdf

IES. (2024). *Functional reasoning.* Retrieved August 3, 2024, from https://ies.ed.gov/ncee/edlabs/infographics/pdf/REL_SE_Functional_Reasoning_Part_of_the_Development_of_Mathematical_Reasoning.pdf

McCallum, W. G., Connally, E., & Hughes-Hallett, D. (2010). *Algebra: Form and function.* Retrieved August 10, 2024, from https://yourknow.com/uploads/books/Algebra_Form_and_Function.pdf

Murdock, J., Kamischke, E., & Kamischke, E. (2014). *Discovering Algebra.* Kendall Hunt Publishing Company.

Murdock, J., Kamischke, E., Kamischke, E., & Harris, P. (2017). *Discovering advanced algebra: An investigative approach.* Kendall Hunt Publishing Company.

CHAPTER 7

Ball, D. L., Thames, M. H., & Phelps, G. (2008). Content knowledge for teaching: What makes it special? *Journal of Teacher Education, 59*(5), 389–407.

Boaler, J. (2022). *Mathematical mindsets.* Jossey Bass.

Carpenter, et al. (2015). *Children's mathematics: Cognitively guided instruction.* Heinemann

Dowker, A. (1992). Computational estimation strategies of professional mathematicians. *Journal for Research in Mathematics Education, 23*(1), 44–55.

Dweck, C. S. (2006). *Mindset: The new psychology of success.* Random House.

Empson, S. B., & Levi, L. (2011). *Extending children's mathematics: Fractions and decimals.* Heinemann.

Fosnot, C. T., & Dolk, M. (2001). *Young mathematicians at work: Constructing number sense, addition, & subtraction.* Heinemann Educational Books.

Fosnot, C. T., & Uittenbogaard, W. (2007). *Minilessons for extending multiplication and division.* Contexts for Learning.

Freudenthal, H. (1987). *Mathematics starting and staying in reality* [Conference session]. Proceedings of the USCMP Conference on Mathematics Education on Development in School Mathematics around the World (pp. 279–295). NCTM.

Gravemeijer, K., & Terwel, J. (2000). Hans Freudenthal: A mathematician on didactics and curriculum theory. *Journal of Curriculum Studies, 32*(6), 777–796. http//doi.org/10.1080/00220270050167170

Hartnett, J. (2015). Teaching computations in primary school without traditional written algorithms. In M. Marshman, V. Gieger, & A. Bennison (Eds.). Mathematics education in the margins. (Proceedings of the 38th annual conference of the Mathematics Education Research group of Australasia), pp. 285–292. Sunshine Coast: MERGA.

Hurst, C., & Huntley, R. (2018). Algorithms . . . Alcatraz: Are children prisoners of process? *International Journal for Mathematics Teaching and Learning, 19*(1), 47–68.

Lambert, R. (2024). *Rethinking disability and mathematics: A UDL Math classroom guide for grades K–8.* Corwin.

NCTM. (2014). *Principles to actions: Ensuring mathematical success for all.* Author.

Vakharia, V. (2024). *Math Therapy™: 5 steps to help your students overcome math trauma and build a better relationship with math.* Corwin.

van den Heuvel-Panhuizen, M., & Wijers, M. (2005). Mathematics standards and curricula in the Netherlands. *ZDM, 37*(4), 287–307.

Index

A Sage Company

CORWIN HAS ONE MISSION: to enhance education through intentional professional learning.

We build long-term relationships with our authors, educators, clients, and associations who partner with us to develop and continuously improve the best evidence-based practices that establish and support lifelong learning.

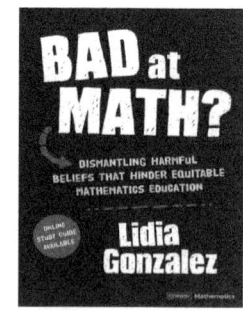

**JENNIFER M. BAY-WILLIAMS,
JOHN J. SANGIOVANNI, ROSALBA SERRANO,
SHERRI MARTINIE, JENNIFER SUH,
C. DAVID WALTERS, SUSIE KATT**

Because fluency is so much more than
basic facts and algorithms.
Grades K–8

**MARIA DEL ROSARIO
ZAVALA,
JULIA MARIA AGUIRRE**

Discover innovative equity-
based culturally responsive
mathematics instruction that
unlocks the mathematical
heart of each student.
Grades K–8

**IVANNIA SOTO,
THEODORE RUIZ SAGUN,
MICHAEL BEIERSDORF**

Focus on the literacy
opportunities that multilingual
students can achieve when
language scaffolds are
taught alongside rigorous
math and science content.
Grades K–8

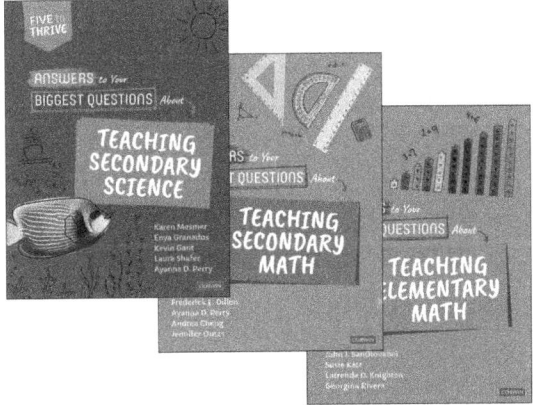

**JOHN J. SANGIOVANNI, SUSIE KATT,
LATRENDA D. KNIGHTEN, GEORGINA RIVERA,
FREDERICK L. DILLON, AYANNA D. PERRY,
ANDREA CHENG, JENNIFER OUTZS, KAREN MESMER,
ENYA GRANDOS, KEVIN GANT, LAURA SHAFER**

Actionable answers to your most pressing
questions about teaching elementary math,
secondary math, and secondary science.

Elementary, Secondary

**CHRISTA JACKSON, KRISTIN L. COOK,
SARAH B. BUSH,
MARGARET MOHR-SCHROEDER,
CATHRINE MAIORCA, THOMAS ROBERTS**

Help educators create integrated STEM
learning experiences that are inclusive for all
students and allow them to experience STEM
as scientists, innovators, mathematicians,
creators, engineers, and technology experts!

Grades PreK–5 and Grades 6–12

CORWIN